BrightRED Study Guide

Curriculum for Excellence

N5

BIOLOGY

Margaret Cook and Fred Thornhill

First published in 2013 by:
Bright Red Publishing Ltd
1 Torphichen Street
Edinburgh
EH3 8HX

New edition published in 2017 following SQA course changes

A CIP record for this book is available from the British Library

ISBN 978-1-84948-312-4

With thanks to:
PDQ Digital Media Solutions Ltd, Bungay (layout and illustrations)
and Dr Anna Clark (copy-edit)
Cover design and series book design by Caleb Rutherford – e i d e t i c

Acknowledgements
Every effort has been made to seek all copyright-holders. If any have been overlooked, then Bright Red Publishing will be delighted to make the necessary arrangements.

Permission has been sought from all relevant copyright holders and Bright Red Publishing are grateful for the use of the following:

Victor M. Vicente Selvas/Public Domain (p 12); LUMEN – Loyola University Medical Education Network (p 16); International Rice Research Institute (IRRI)/Creative Commons (CC BY 2.0)[1] (p 23); 18percentgrey/Shutterstock.com (pp 31 and 32); stefanolunardi/Shutterstock.com (p 31); motorolka/Shutterstock.com (p 32); luchschen/Shutterstock.com (p 32); 2 photos by Dmitry Lobanov/Shutterstock.com (p 50); Konstantins Visnevskis/Shutterstock.com (p 50); S.john/Shutterstock.com (p 51); MichaelTaylor/Shutterstock.com (p 55); Madlen/Shutterstock.com (p 56); Richard Parker/Creative Commons (CC BY 2.0)[1] (p 56); doglikehorse/Shutterstock.com (p 56); BritishEmpire/Creative Commons (CC BY-SA 3.0)[3] (p 56); Ectelion7/Creative Commons (CC BY-SA 4.0)[8] (p 56) ; Dan Kosmayer/123rf.com (p 56); nrt/Shutterstock.com (p 56); Department of Statistics and Actuarial Science, Simon Fraser University (p 57); Jing Jane Wang (p 60); AJ Cann/Creative Commons (CC BY-SA 2.0)[2] (p 65); D. Kucharski K. Kucharska/Shutterstock.com (p 65); Nixx Photography/Shutterstock.com (p 66); Andrea Danti (p 68); Jacopo Werther/Creative Commons (CC BY-SA 2.0)[2] (p 75); André Karwath/Creative Commons (CC BY-SA 2.5)[5] (p 75); Charles Haynes/Creative Commons (CC BY-SA 2.0)[2] (p 75); Tony Wills/Creative Commons (CC BY 2.5)[5] (p 75); Fred Thornhill (p 75); Ian Rentoul/Shutterstock.com (p 78); BirdPhotos.com/Creative Commons (CC BY 3.0)[6] (p 78); Aaron Gustafson/Creative Commons (CC BY-SA 2.0)[2] (p 78); cogito ergo imago/Creative Commons (CC BY-SA 2.0)[2] (p 78); tap78/Stock-Xchnge (p 80); smart.art/Shutterstock.com (p 80); Radu Razvan/Shutterstock.com (p 80); Brandon Blinkenberg/Shutterstock.com (p 81); James K. Lindsey/Creative Commons (CC BY-SA 2.5)[5] (p 86); Fir0002/Flagstaffotos/Creative Commons (GFDL v1.2)[7] (p 86); Quartl/Creative Commons (CC BY-SA 3.0)[3] (p 86); Alvesgaspar/Creative Commons (CC BY-SA 3.0)[3] (p 86); Jacopo Werther/Creative Commons (CC BY 2.0)[1] (p 86); Dalavich/Creative Commons (CC BY-SA 3.0)[3] (p 86); Darkone/Creative Commons (CC BY-SA 2.5)[5] (p 86); Guttorm Flatabø/Creative Commons (CC BY-SA 3.0)[3] (p 86); Vaikoovery/Creative Commons (CC BY 3.0)[6] (p 86); Rob Hille/Public Domain (p 86); Leon Brooks/Public Domain (p 86); Dirk Ercken/Shutterstock.com (p 88); guentermanaus/Shutterstock.com (p 89); BastienM/Public Domain (p 89); Amada44/Creative Commons (CC BY 2.0)[1] (p 90); Lairich Rig/Creative Commons (CC BY-SA 2.0)[2] (p 90); Andrea Pokrzywinski/Creative Commons (CC BY 2.0)[1] (p 90); H. Zell/Creative Commons (CC BY-SA 3.0)[3] (p 90); Two photos by Dave Huth/Creative Commons (CC BY 2.0)[1] (p 91); XenonX3/Public Domain (p 91); Kim Fleming (p 91); James Niland/Creative Commons (CC BY 2.0)[1] (p 91); Becky Stares/Shutterstock.com (p 91); CWA Studios/Shutterstock.com (p 101); US Fish & Wildlife photo/Public Domain (p 102); Arne and Bent Larsen/Creative Commons (CC BY-SA 2.5)[5] (p 102); Natural History Museum of London/Creative Commons (CC BY-SA 3.0)[3] (p 102); Maximilian Paradiz/Creative Commons (CC BY 2.0)[1] (p 102); Donald Macauley/Creative Commons (CC BY-SA 2.0)[2] (p 102); Stubblefield Photography (p 104); Daniel Alvarez (p 104); Ryan M. Bolton (p 104); Zouavman Le Zouave/Creative Commons (CC BY-SA 3.0)[3] (p 105); Rosser1954/Public Domain (p 105); Rob & Sue Brookes (p 105); nottsexminer/Creative Commons (CC BY-SA 2.0)[2] (p 105).

[1] (CC BY 2.0) http://creativecommons.org/licenses/by/2.0/
[2] (CC BY-SA 2.0) http://creativecommons.org/licenses/by-sa/2.0/
[3] (CC BY-SA 3.0) http://creativecommons.org/licenses/by-sa/3.0/
[4] (CC BY-ND 2.0) http://creativecommons.org/licenses/by-nd/2.0/
[5] (CC BY-SA 2.5) http://creativecommons.org/licenses/by-sa/2.5/
[6] (CC BY 3.0) http://creativecommons.org/licenses/by/3.0/
[7] (GFDL v1.2) http://www.gnu.org/licenses/fdl-1.2.html#SEC1
[8] (CC BY-SA 4.0) https://creativecommons.org/licenses/by-sa/4.0/deed.en

Printed in the UK.

CONTENTS

BRIGHTRED STUDY GUIDE: NATIONAL 5 BIOLOGY

1 CELL BIOLOGY

2 MULTICELLULAR ORGANISMS

3 LIFE ON EARTH

GLOSSARY

INTRODUCING NATIONAL 5 BIOLOGY

During this course we hope you will develop and apply skills for learning, skills for life and skills for work.

Biology is the study of living organisms so it is relevant to everybody. It plays a crucial role in our everyday existence, and is an increasingly important subject in the modern world. Advances in biological technologies have made the subject more exciting and more relevant than ever.

THE NATIONAL 5 BIOLOGY COURSE

National 5 Biology encourages you to become:

- a more confident learner
- a responsible citizen with an informed and ethical view of complex issues, through the study of relevant areas of biology such as health, environment and sustainability
- someone who thinks analytically, creatively and independently, and is able to make reasoned evaluations.

The National 5 course provides opportunities for you to acquire knowledge and skills relevant to current biological topics. The course covers major areas of biology, ranging from the study of unicellular organisms to the complex relationships between organisms in an ecosystem. Key areas focus on cells and cellular processes, leading to an understanding of the importance of cells and their roles. Body systems, reproduction and inheritance are investigated to broaden your understanding of living organisms. The comparison of the processes in multicellular plants and animals enables you to investigate increasing levels of complexity of these organisms. Key areas of biodiversity and interdependence are covered, along with the processes leading to evolution, and food security.

COURSE CONTENT

You will develop skills of scientific inquiry, investigation and analytical thinking, along with the required knowledge and understanding.

You will also research issues and communicate information related to your findings, developing skills of scientific literacy.

Topic 1 - Cell biology

The key areas covered are: cell structure; transport across cell membranes; DNA and the production of proteins; proteins; genetic engineering; and respiration.

Topic 2 - Multicellular organisms

The key areas covered are: producing new cells; control and communication; reproduction, variation and inheritance; transport systems – plants; transport systems – animals; and absorption of materials .

Topic 3 - Life on Earth

The key areas covered are: ecosystems; distribution of organisms; photo synthesis; energy in ecosystems; food production; and evolution in species.

The range of skills involved in the course are:

- demonstrating knowledge and understanding of biology by describing information, explaining and linking knowledge

contd

- applying biological knowledge to new situations, interpreting information and solving problems
- planning, designing and safely carrying out investigations and experiments to test hypotheses or to show particular effects
- selecting information from a variety of sources
- presenting information in a variety of forms
- processing information (for example using calculations and units)
- making predictions and generalisations based on evidence and information
- justifying conclusions that are supported by evidence
- suggesting improvements to investigations and experiments
- communicating findings and information.

THE EXTERNAL ASSESSMENT

At the end of the course you will be assessed externally by two components.

Component 1 – Question Paper (80% of total mark)

This is made up of a 2½-hour question paper in which:

- 25 marks are allocated to an objective test
- 75 marks are allocated to the written paper, which includes questions requiring a mixture of short (restricted) and extended answers.

Many of the marks will be given for applying knowledge and understanding. The other marks will be given for applying scientific inquiry, analytical thinking and problem-solving skills.

The question paper will be written and marked by the Scottish Qualifications Authority (SQA).

Component 2 – Assignment (20% of total mark)

The assignment will be an in-depth study of a biology topic you have chosen. There will be 20 marks awarded for the assignment and the majority of these will be awarded for applying scientific inquiry and analytical thinking skills. The other marks will be awarded for applying knowledge and understanding relating to the topic.

The assignment is carried out under controlled conditions and is externally marked by SQA. To prepare for the controlled assessment you will choose, research and investigate an appropriate topic. Your research involves gathering data or information from an experiment or fieldwork as well as from the internet or literature.

During the assessment you will present evidence of:

- an aim you have decided upon
- biological knowledge and understanding relating to your chosen topic
- a brief description of an experiment or fieldwork you have carried out, along with the results
- data or information from the internet or literature to compare with your experimental data
- a reasoned conclusion.

ONLINE

This book is supported by the Bright Red Digital Zone. Visit www.brightredbooks. net/N5Biology and discover a world of tests, videos, activities and more!

CELL BIOLOGY

CELL STRUCTURE 1

CELL ORGANELLES IN PLANT AND ANIMAL CELLS

Powerful microscopes, called **electron microscopes**, show that **cells** are much more complex than they appear through light microscopes, such as those used in schools. The detailed structure of cells that is revealed using electron microscopes is known as the cell ultrastructure. The small structures found inside cells are known as **organelles**. Each type of organelle has a specific function, although not all organelles are found in all cells.

The following diagrams show the organelles commonly found in plant and animal cells.

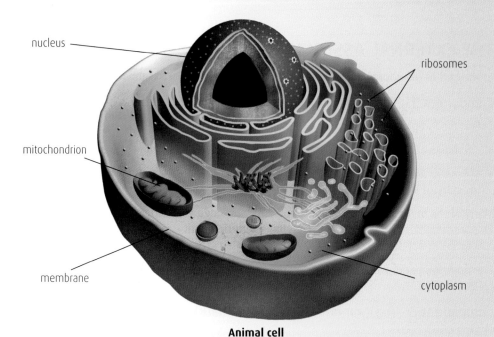

Animal cell

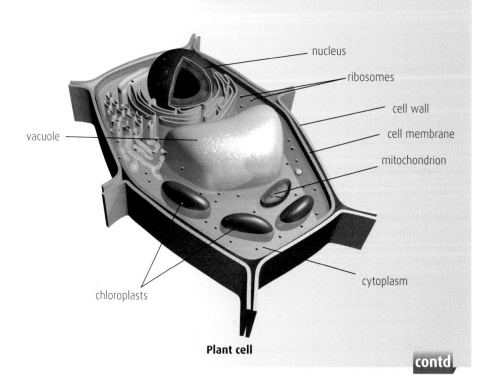

Plant cell

contd

VIDEO LINK

Watch the 'Plant and Animal Cell Overview' video for a more detailed look inside plant and animal cells at www.brightredbooks.net/N5Biology

DON'T FORGET

Some diagrams show cell parts in 3-D while others show them in 2-D as 'cut-through' diagrams. It is useful to recognise both forms.

All cells have a **cell membrane** which forms the boundary of the cell. The cell is filled with a jelly-like material called **cytoplasm** in which chemical reactions take place. The cytoplasm contains the organelles.

The following table describes the various parts of cells and their functions, as well as stating where they are found.

Part of cell	Function	Organisms that have these
Nucleus	This contains the cell **chromosomes** which are made of **DNA**. These hold the genetic information which controls cell activities.	Plants and animals
Cell membrane	This consists of **phospholipid** and **protein** molecules. It is **selectively permeable** and controls the entry and exit of materials in and out of the cell.	Plants and animals
Mitochondrion (Mitochondria – plural)	This is the site of **aerobic respiration**. Mitochondria are found in greater abundance in cells with high energy demands, such as muscle cells.	Plants and animals
Ribosome	This is the site of protein synthesis. Ribosomes are found in the cytoplasm or attached to tubular structures in the cell.	Plants and animals
Chloroplast	This is the site of **photosynthesis**. Chloroplasts contain the green pigment **chlorophyll**, which absorbs light energy.	Plants
Vacuole	This contains cell sap, which is a solution of salts and sugars. It is important in maintaining the shape of the cell.	Plants
Cell wall	This gives support to the cell and structure to plant tissue. It is made of a carbohydrate called **cellulose**.	Plants

VARIATION IN PLANT AND ANIMAL CELLS

Multicellular organisms are made up of a large number of cells which work together to allow the organism to function as a unit. The previous diagrams show the contents of typical plant and animal cells. Most cells are, however, adapted to perform a particular function within an organism. Therefore, there is variation between cells. This will be explored further in a later chapter.

 ONLINE TEST

Test how well you've learned about the structure of plant and animal cells by taking the 'Cell structure' test online at www.brightredbooks.net/N5Biology

 THINGS TO DO AND THINK ABOUT

Look at the following diagram of a cell and complete the table to identify the various organelles and their functions.

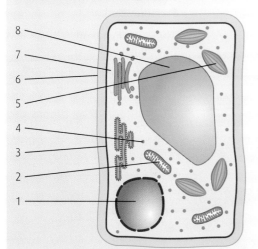

Part	Name	Function
1		
2		
3		
4		
5		
6		
7		
8		

CELL STRUCTURE 2

CELLS OF DIFFERENT TYPES OF ORGANISMS

Unicellular organisms

Unicellular organisms exist as single cells. Each of the cells must contain all of the organelles necessary for survival. Unicellular organisms can be plant-like or animal-like. Some examples and their ultrastructure are shown below.

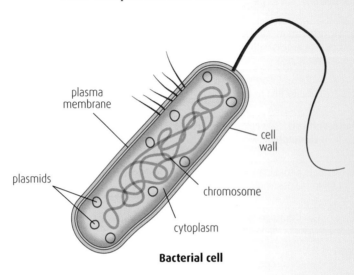

Bacterial cell

Bacterial cells

Bacterial cells differ from plant and animal cells in several ways. They do not have organelles. Although they do have a cell wall, it is not made of cellulose. They do not have a nucleus to contain their genetic material. Instead they have one large loop of chromosomal material and several much smaller rings known as **plasmids**. Due to their size, plasmids can easily be removed from and reinserted into bacterial cells. This technique is important in **genetic engineering** and will be explored in a later chapter.

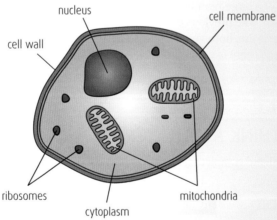

Fungal cell

Fungal cells

Fungal cells show some similarities and some differences when compared to plant and animal cells. Like bacteria, their cell wall is made of a carbohydrate which is not cellulose. One of the most familiar fungal cells is yeast, which is used in the process of **fermentation**.

There are many other types of unicellular organisms such as *Amoeba*, *Euglena*, *Paramecium* and algae.

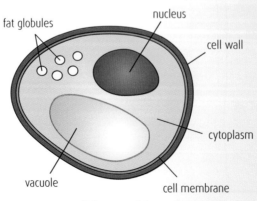

Yeast cell (a type of fungus)

VIDEO LINK

Watch the clips of *Amoeba*, *Paramecium* and *Euglena* at www.brightredbooks.net/N5Biology

CELL SIZES

Cells vary greatly in size. As they are all microscopic, the scales of measurement normally used (metres, centimetres, millimetres) are too large to measure them. Therefore, a smaller measurement known as a micrometre is used to measure cells. There are 1000 micrometres (μm) to 1 millimetre (mm).

Fungal and bacterial cells are much smaller than typical plant and animal cells.

The table below shows the sizes of some cells.

Type of cell	Approximate size (μm)
E. coli – a bacterium	2
Yeast	3
Human red blood cell	9
Typical animal cell	10–30
Typical plant cell	10–100
Small amoeba	90
Human egg	100

DON'T FORGET

Micrometres are also known as microns.

ONLINE TEST

Test how well you've learned about the structure of plant and animal cells by taking the 'Cell structure' test online at www.brightredbooks.net/N5Biology

THINGS TO DO AND THINK ABOUT

1   Look at the following diagram of a cell and complete the table to identify the various organelles and their functions.

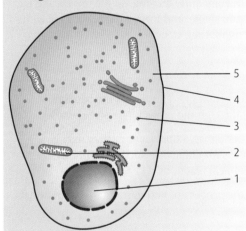

Part	Name	Function
1		
2		
3		
4		
5		

2   Compare this cell with the one shown on p6. Is this cell from a plant or an animal? Give evidence to support your answer.

3   You should be able to compare relative sizes of cells and convert sizes between micrometres (μm) and millimetres (mm):

- A measurement in mm can be converted to μm **by multiplying by 1000.**

- A measurement in μm can be converted to mm **by dividing by 1000**.

Complete the following table.

mm	μm
0·33	
	85
0·76	

TRANSPORT ACROSS CELL MEMBRANES

THE STRUCTURE OF THE CELL MEMBRANE

The cell membrane consists of **phospholipids** (fats) and proteins arranged as shown in the diagram.

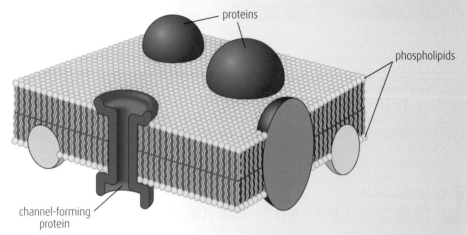

Structure of the cell membrane

The phospholipids form a double layer and are constantly moving, so the membrane is described as 'fluid'. The proteins vary in size and shape, and form a patchy mosaic within the phospholipid bilayer. For this reason, the membrane structure is referred to as the **fluid mosaic model**.

The proteins have a variety of functions:

- Some form channels, creating pores in the membrane. This allows the passage of molecules that are small enough to travel through the pores by passive means.

- Some proteins are partially embedded in the phospholipid layer or they may lie on its surface. Some of these act as receptors for chemicals such as **hormones** or **antibodies**. Others act as **enzymes** for chemical reactions.

- Other proteins act as carrier proteins for the process of **active transport** of materials across the membrane.

The cell membrane is described as being selectively permeable. This term refers to the property of allowing some molecules to pass from one side to the other, while other molecules are unable to pass through. Smaller molecules (such as water, oxygen and carbon dioxide) move freely through the cell membrane, slightly larger ones (for example **amino acids**, glucose and urea) move more slowly through it, but large ones (like proteins and **starch**) are unable to pass through.

PASSIVE METHODS OF TRANSPORT

Passive transport of materials does not involve energy expenditure by the organism for the process to take place. It relies on the fact that substances are found in different concentrations in different places. With passive transport, substances always move from a higher to a lower concentration. This difference in concentration is known as a **concentration gradient**.

contd

Diffusion

The definition of **diffusion** is the movement of molecules or ions from a region of higher concentration to a region of lower concentration.

Diffusion allows substances to move within a cell from one area to another, as well as allowing materials to enter and leave a cell. Diffusion continues until the concentration of the substance is even throughout.

Diffusion is important to organisms in many ways. For example, oxygen and glucose enter cells by diffusion for the process of aerobic respiration. The removal of waste products, such as carbon dioxide or urea, from cells also happens by diffusion.

Osmosis

The definition of **osmosis** is the movement of water molecules from a region of higher water concentration to a region of lower water concentration through a selectively permeable membrane.

Osmosis is a special type of diffusion and only involves the movement of water molecules.

If there is a difference in water concentration between the fluids inside and outside of the cell, then water will either enter or leave the cell, moving from the higher to the lower water concentration. If the water concentration is equal on either side of the membrane, water molecules will still move from one side to the other, but in equal proportions. This is not osmosis, since osmosis (like diffusion) requires a concentration gradient to occur.

Osmosis is a very important cellular process as all cells require water to survive.

ACTIVE METHODS OF TRANSPORT

The definition of **active transport** is the movement of molecules and ions across a cell membrane against a concentration gradient.

Unlike diffusion and osmosis, molecules or ions do not move by themselves in active transport. Instead, they are moved by carrier proteins, which are a component of membranes. This process requires energy because molecules are being moved against the concentration gradient; that is from lower to higher concentration. The energy is provided by **ATP** (adenosine triphosphate), which is produced during **respiration**. Therefore, anything that limits respiration will also limit the rate of active transport.

One example of the importance of active transport in animal cells is the reabsorption of glucose from the kidney tubules back into the bloodstream, ensuring that glucose is not passed out in the urine. An example of active transport in plants can be found in the fact that some seaweeds and corals can accumulate iodine from sea water in their cells to a concentration hundreds of times greater than that in the surrounding water.

VIDEO LINK

Have a look at the video on diffusion at www.brightredbooks.net/N5Biology

VIDEO LINK

Check out the simulation of osmosis at www.brightredbooks.net/N5Biology

DON'T FORGET

Diffusion and osmosis are two types of passive transport. Neither process requires energy expenditure by the organism

DON'T FORGET

Active transport goes against a concentration gradient and requires energy expenditure by the organism

ONLINE TEST

Test yourself on transport across cell membranes at www.brightredbooks.net/N5Biology

THINGS TO DO AND THINK ABOUT

The table shows the concentration of some substances in a cell and in the environment surrounding it. Identify the method of transport that would in each case move molecules of each substance into the cell.

Substance	Concentration in cell (%)	Concentration surrounding cell (%)
Glucose	5	10
Water	90	98
Oxygen	16	20
Carbon dioxide	3	2

EFFECTS OF OSMOSIS ON PLANT AND ANIMAL CELLS

Animal cells and plant cells react differently to the intake or loss of water. This is due to the differences in their structures.

OSMOSIS AND PLANT CELLS

As plant cells have a vacuole for water storage, any water gain or loss from a cell affects the size of the vacuole. This has an overall effect on the rest of the cell, as shown in the diagram below.

A plant cell in a solution which has an equal concentration to the solution in the vacuole will experience no net gain or loss of water. The cell will remain unchanged.

However, if the solution surrounding the plant cell has a water concentration greater than that in the cell, water passes into the cell from the surroundings, causing the vacuole to swell and press outwards against the cytoplasm. This, in turn, causes the cytoplasm to push against the cell membrane and cell wall. The cell wall is fairly rigid and strong enough to resist this pressure, preventing the cell from bursting. In this state the cell is described as **turgid**.

On the other hand, if the solution surrounding the cell has a water concentration lower than that in the cell, water passes out of the cell to its surroundings, causing the vacuole to shrink and pulling the cytoplasm inwards. This drags the cell membrane inwards too, but the cell wall remains in place. A gap develops between the membrane and the cell wall and, as the cell wall is fully permeable, the surrounding solution flows in to fill this gap. In this state, the cell is described as being **plasmolysed**.

Plant cells in various solutions

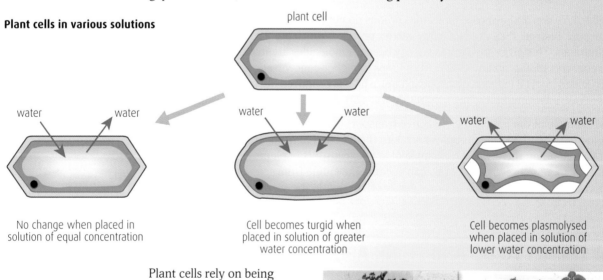

No change when placed in solution of equal concentration

Cell becomes turgid when placed in solution of greater water concentration

Cell becomes plasmolysed when placed in solution of lower water concentration

Plant cells rely on being turgid to maintain their shape. Leaves can be seen to wilt if the cells are lacking the water content to keep them turgid. The cells become **flaccid** and no longer have enough pressure inside to maintain the upright position of the leaf. They can, however, recover if the plant is watered.

Wilted and recovered plants

OSMOSIS AND ANIMAL CELLS

Animal cells react differently to plant cells in terms of water gain or loss, due to the fact that they have neither a cell wall nor a vacuole.

An animal cell in a solution which has an equal concentration to the solution in the cell will experience no net gain or loss of water. The cell will remain unchanged.

However, if the solution surrounding the animal cell has a water concentration greater than that in the cell, water passes into the cell from the surroundings, causing the cell to swell up. As there is no cell wall to resist the continued stretching of the membrane, the cell eventually bursts.

On the other hand, if the solution surrounding the cell has a water concentration lower than that in the cell, water passes out of the cell to its surroundings, causing the cell to shrink and shrivel up.

As red blood cells are surrounded by a solution (blood plasma), it is important that the body has mechanisms in place to prevent the plasma becoming too dilute or too concentrated as blood cells could burst or shrink.

 DON'T FORGET

When referring to an increase in *water concentration*, this means that the concentration of the dissolved substance in the solution decreases. Therefore, water always moves by osmosis from a weak solution to a strong solution.

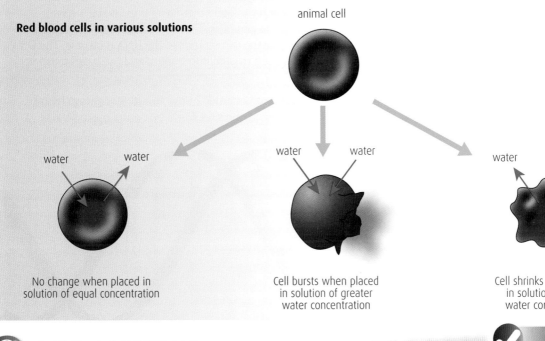

Red blood cells in various solutions

animal cell

water   water

water   water

water   water

No change when placed in solution of equal concentration

Cell bursts when placed in solution of greater water concentration

Cell shrinks when placed in solution of lower water concentration

THINGS TO DO AND THINK ABOUT

Plants store the carbohydrate they produce during photosynthesis in the form of starch. Starch is a large molecule and is insoluble in water. Why is it important for the plant not to store carbohydrate as small soluble molecules of sugar?

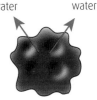 **ONLINE TEST**

Test yourself on osmosis at www.brightredbooks.net/N5Biology

DNA AND THE PRODUCTION OF PROTEINS

Double helix of DNA

THE STRUCTURE OF DNA

Variation exists among the members of any given species. This is mainly due to differences in the information held in the genetic material and stored in the nucleus of their cells. The genetic information is in the form of chromosomes.

The nucleus of every cell contains thread-like structures known as chromosomes. They are made from a chemical called DNA (deoxyribonucleic acid). This is a double-stranded molecule twisted into a spiral known as a **double helix**. DNA carries the **genetic code** for making proteins. Each chromosome contains the instructions to make hundreds of different types of protein.

Each strand of the DNA molecule is made from a chain of subunits. Each subunit contains a type of base: Adenine (A), Thymine (T), Guanine (G) or Cytosine (C). The letters represent the names of the bases and each of the bases is a different shape. The shape of each base is complementary to one other base; thus, A and T have complementary shapes, and G and C are complementary to each other. The bases are linked by weak bonds, holding together the two strands in the DNA molecule. The DNA double helix is like a ladder that has been twisted into a spiral and the bases make up the 'rungs' of the ladder.

In a DNA double helix, therefore, if A is on one side of the strand, T is always found on the opposite strand. Similarly, if G is on one side, C will be on the opposite side (A–T and G–C).

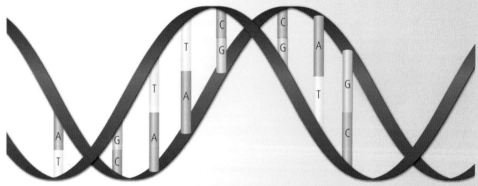

Thymine (Yellow) = T   Guanine (Green) = G
Adenine (Blue) = A   Cytosine (Red) = C

DNA double helix showing base pairing

The sequence of the bases in a strand of DNA determines the sequence of amino acids which are joined together to make a protein. Different sequences of bases make different proteins. A chromosome is made up of hundreds of **genes**, each coding for a particular protein.

PROTEIN SYNTHESIS

All proteins are made of subunits called amino acids. These are chemicals which contain the elements carbon, hydrogen, oxygen and nitrogen. There are about 20 different amino acids. The sequence of these amino acids in a protein determines its structure and function.

A section of DNA which codes for a protein is called a gene. It can be hundreds of bases in length. The bases form groups of three, called triplets. Each triplet carries the code

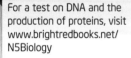

ONLINE TEST

For a test on DNA and the production of proteins, visit www.brightredbooks.net/N5Biology

contd

for a particular amino acid. The amino acids must be arranged in the correct sequence to make a particular protein. The order of the triplets of bases determines this.

Using the DNA as a template, molecules called messenger **RNA** (mRNA) are made in the nucleus. Base pairing ensures that these are complementary to the section of the DNA (a gene). The mRNA then leaves the nucleus and travels to a ribosome.

A protein can only be assembled within a ribosome. Ribosomes are found either individually or attached to a system of tubes in the cytoplasm. The ribosome 'reads' the sequence of bases on the mRNA, ensuring that the correct amino acid is carried into place as it moves along the stand of mRNA.

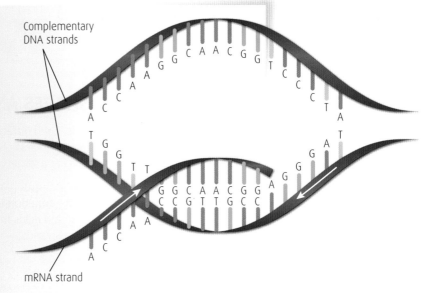

mRNA synthesis from a DNA molecule

As each amino acid is brought into line, it is joined to its neighbour by a chemical bond until a long chain is formed. These chains are then shaped into the form required for that particular protein.

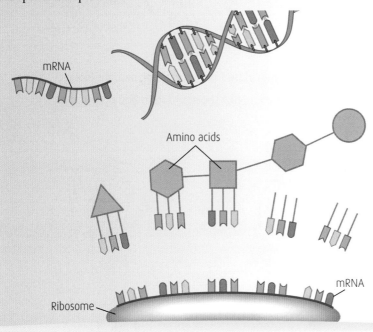

Sequence of events in translation of the code into a protein

It can be seen that the sequence of the bases in the DNA in the nucleus determines the final structure of the protein made in the ribosome.

VIDEO LINK

Watch the simulation of DNA replication at www.brightredbooks.net/N5Biology

VIDEO LINK

Watch the animation about protein synthesis at www.brightredbooks.net/N5Biology

DON'T FORGET

The order of the bases in the DNA determines the sequence of amino acids in a protein. This is brought about by base pairing in the production of mRNA and at the ribosome where the amino acids are assembled into a protein.

THINGS TO DO AND THINK ABOUT

1 What are the units that make up a protein?

2 How is the sequence of amino acids in a protein determined?

3 Why is the sequence of amino acids in a protein of importance?

PROTEINS 1

DIFFERENT PROTEIN SHAPES AND FUNCTIONS

All proteins are made of amino acids. Each different protein is made of a different sequence of amino acids. Once the amino acids have been assembled into a long chain, they are folded in various ways to make a variety of shapes. Each particular protein has its own unique shape.

The shape of a protein is related to the function that it carries out. The shape is achieved by folding the strands of amino acids in certain ways after they leave the ribosome. Some are twisted into long strands, while others are wound into a roughly spherical shape, like a ball of wool.

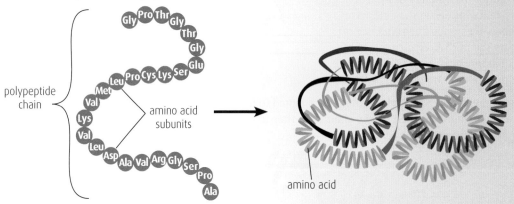

Formation of the enzyme amylase from several polypeptide chains

Some proteins have a **structural** function, like those found in body tissues such as muscles, ligaments and tendons.

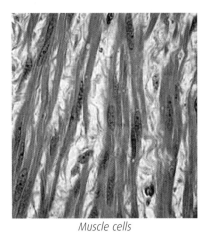

Muscle cells

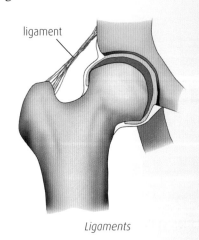

Ligaments

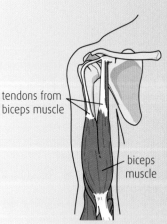

Tendons

Structural proteins

- Muscle cells contain long thin fibres of protein, allowing the cells to contract and relax.

- Ligaments are made of very strong, slightly elastic, fibrous proteins which hold bone to bone at a joint, preventing dislocation.

- Tendons have very tough fibres of protein which can withstand the pulling force of muscle on bone without stretching.

Other proteins, like antibodies, hormones and enzymes, are folded into a range of very specific shapes to suit their particular function.

contd

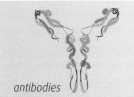
antibodies

hormones

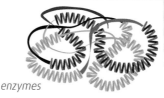
enzymes

Protein shape is related to function

- Antibodies are proteins that are produced by the body's white blood cells in response to invasion by pathogens. Antibodies are Y-shaped and attach to the pathogen to prevent it attacking cells.

- Hormones are chemical messengers which target specific tissues to cause a particular response. There are a great many types.

- Enzymes are proteins folded in such a way that they have an **active site** which binds to another substance to bring about a reaction.

Some proteins are found on the surface of cells and act as receptors, allowing, for example, a hormone to attach to the receptor and cause a change in the cell.

ENZYMES

All enzymes are made of protein. They are made inside living cells by the process of protein synthesis. Each enzyme has a particular shape which determines its function.

Enzymes are biological **catalysts** and speed up chemical reactions. They:

- lower the activation energy required to start the chemical reaction
- speed up the rate of the chemical reaction
- remain unchanged after bringing the reaction about.

Enzymes are involved in all biochemical reactions. Living organisms exist at relatively low temperatures and are unable to tolerate the high temperatures required to bring about rapid chemical reactions. Enzymes allow essential reactions to take place at lower temperatures. Without the use of enzymes, the chemical reactions in cells would take place too slowly to sustain life. Since enzymes are not used up in a reaction, they can be used over and over again.

Enzymes can carry out two types of reactions:

1 **degradation** – breaking down a large molecule into two or more smaller ones

2 **synthesis** – building up two or more molecules to a larger one.

The substance with which the enzyme reacts is known as the **substrate** and the substance produced at the end of the reaction is called the **product**. Therefore, in an enzyme-controlled reaction:

$$\text{Substrate} \xrightarrow{\text{Enzyme}} \text{Product}$$

On the surface of an enzyme molecule is an area known as the active site. This is the area that the substrate molecule(s) binds to in order for the chemical reaction to take place. The shape of the active site is determined by the folding of the chains of amino acids which make up the enzyme protein. The active site and the substrate have complementary shapes, so that the substrate can fit into the active site of the enzyme.

THINGS TO DO AND THINK ABOUT

1 Try to find examples of different categories of proteins. Can you name two hormones and two enzymes?

2 What is haemoglobin? Where is it found?

DON'T FORGET

Proteins are three-dimensional structures and each type varies according to the sequence of its amino acids and the shape it is folded into.

DON'T FORGET

Enzymes are folded into particular shapes, so that their active site only fits one substrate. Enzymes can be involved in degradation, as well as synthesis of molecules.

ONLINE TEST

Test yourself on proteins and enzymes at www.brightredbooks.net/N5Biology

substrate

active site

enzyme

Enzyme with substrate that fits the active site

PROTEINS 2

HOW ENZYMES WORK

Each enzyme is only able to act on one type of substance, due to the shape of its active site. It is said to be **specific** in its reaction. A substrate with a different shape will not fit into the active site, so will not be affected by the enzyme.

Enzymes and substrates are described as operating like a **lock and key**. A key only fits a specific lock and can change it from one state to another (locked to unlocked); the key remains unchanged and can, therefore, be used again. Similarly, an enzyme only fits a specific substrate and changes it into some sort of product, without being used up in the process.

An enzyme molecule binds with a substrate molecule to form an enzyme–substrate complex. This allows the chemical reaction to take place. At the end of the reaction, the product(s) detach from the active site and leave the enzyme. This means that the enzyme is now free to combine with another substrate molecule and catalyse the reaction again.

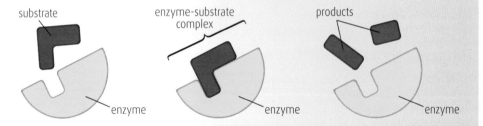

Lock and key hypothesis of enzyme action

ENZYME ACTION

Enzymes either break down or build up molecules.

Degradation (breaking down)

Some enzymes break down complex molecules to simpler ones. All of the enzymes involved in the process of digesting our food carry out degradation reactions, converting large insoluble molecules into smaller soluble ones which can be absorbed into the blood.

Examples of degradation enzymes are shown in the following table.

Substrate	Enzyme	Product
Protein	Pepsin	Peptides, then amino acids
Starch	Amylase	Maltose sugar
Fat	Lipase	Glycerol and fatty acids
Hydrogen peroxide	Catalase	Water and oxygen

Synthesis (building up)

Some enzymes build up simple molecules into more complex ones. A very important synthesis reaction is carried out by the enzyme phosphorylase. The substrate involved is a form of glucose found in plants, known as glucose-1-phosphate. Phosphorylase links glucose-1-phosphate molecules into long chains, which become molecules of starch. This is of benefit to the plant, enabling it to store food in its cells.

$$\text{Glucose-1-phosphate} \xrightarrow{\text{Phosphorylase}} \text{Starch}$$

FACTORS AFFECTING ENZYME ACTIVITY

The ideal conditions for an enzyme to work are called the optimum conditions. These include a suitable temperature, an appropriate pH and a plentiful supply of substrate. Each enzyme is most active when it is in its optimum conditions. If any of these factors is inadequate, the enzyme will not be able to function at its best.

Effect of temperature on enzyme activity

As all enzymes are made of protein, any factor which affects protein will affect an enzyme. Temperature has a significant effect on proteins and, therefore, on enzymes. Most enzymes function best at about 40°C. This is their optimum temperature.

The graph shows how temperature affects starch digestion by the enzyme amylase over a range of temperatures.

It can be seen that, as the temperature begins to increase towards 40°C, the activity of the enzyme increases. This is because the molecules of the substrate and enzyme move faster with the increased heat energy and collide more often. Therefore, enzyme molecules catalyse the conversion of substrate to product more frequently.

The temperature at which the maximum rate of reaction is reached is known as the optimum temperature.

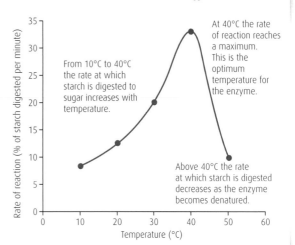

Effect of temperature on enzyme activity

Further increase in temperature causes a decrease in enzyme activity. The increased heat energy begins to break the chemical bonds that hold the protein of the enzyme together, breaking the enzyme apart. The active site loses its shape and the substrate is no longer able to fit into it. In this damaged state, the enzyme is described as denatured.

Denaturation of an enzyme is an irreversible procedure and, as more enzyme molecules are damaged by the heat, enzyme activity will stop altogether.

Effect of pH on enzyme activity

Proteins (and, therefore, enzymes) are affected by the pH of their surroundings. The shape of an enzyme can be altered if the pH is too extreme and at certain pH levels enzymes become denatured.

Each enzyme is most active at a specific pH, although most can operate within a range to either side of that optimum value. This is called the working range of the enzyme. The majority of enzymes work between pH 5 and pH 9 with an optimum of about pH 7, as shown by the enzyme amylase found in saliva.

However, there are exceptions. An example is pepsin, which is secreted by the lining of the stomach. Pepsin is most active in the very acidic conditions found there, which is due to the presence of hydrochloric acid. The optimum pH for pepsin is 2·5.

Another exception is the enzyme arginase, which is found in the liver and is involved in the production of urea. It has an optimum pH of 10, which is highly alkaline.

Effect of substrate supply on enzyme activity

If the substrate is in plentiful supply and all other factors are at their optimum, the enzyme reaction will proceed at a rapid rate. However, as the substrate gets converted to product and is reduced in quantity, the reaction rate will slow. This is due to the fact that there are fewer collisions between the substrate and the enzyme's active site. Fewer enzyme–substrate complexes form, so fewer reactions take place.

DON'T FORGET

A molecule which fits into the active site on an enzyme is called the substrate. A molecule which leaves the active site of the enzyme after the reaction is complete is called the product.

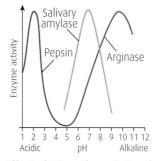

Effect of pH on the activity of different enzymes

THINGS TO DO AND THINK ABOUT

The diagram on page 18 shows the process of degradation of a substance through the 'lock and key' action of an enzyme. Try to draw a set of diagrams to illustrate a synthesis reaction, such as the build up of starch by the enzyme phosphorylase.

GENETIC ENGINEERING

THE TRANSFER OF GENETIC INFORMATION

Genetic material can be transferred from one cell to another, either by natural means or artificially, through the process of genetic engineering.

Scientists have studied the ways in which **viruses** transfer their nucleic acids into host cells. They found that viruses transferred their genetic material into the host so that it became attached to the host's DNA.

Research has shown that some bacteria living in the soil are responsible for causing tumours to develop in plant tissue. Scientists have studied this process to see how it works. They found that the bacterium actually transfers a piece of its own genetic material into the plant root, usually through a wound in the plant tissue. This genetic material is found in one of the bacterial plasmids and is responsible for causing a tumour in the plant root.

Through further research, scientists have developed ways to substitute the genes which cause tumours with other DNA and they are now able to introduce useful genes into bacteria. The bacteria then pass these genes into the plant, allowing the plant to be **genetically modified** in certain ways.

When viruses and bacteria transfer their genetic material into cells, these cells are now able to make proteins which they previously could not. Scientists saw the potential of this and developed ways of using the technique to their advantage.

This technique has been used to create transgenic crops (plants with genes from another organism).

These crops have advantages over the natural varieties, such as increased yields or increased disease resistance, but controversy surrounds their use.

ONLINE

Have a look at the online debate about the banning of transgenic crops in India at www.brightredbooks.net/ N5Biology

GENETIC ENGINEERING

The process of genetic engineering involves taking genetic material from one type of living organism and transferring it into another type of living organism. The organism with the altered genetic make-up is now 'reprogrammed', or transformed, to make different proteins which are useful to human beings.

Microorganisms, such as bacteria and yeast, are often reprogrammed to produce useful substances, including medicines and human proteins such as hormones. There are several advantages to using these single-celled organisms:

- They grow and multiply very quickly.
- Being individually small, they are easy to accommodate.
- They are relatively inexpensive to use.
- They are easier to reprogramme than more advanced organisms.

THE PROCESS OF REPROGRAMMING

The arrangement of the chromosomal material in a bacterium makes it an ideal organism for genetic engineering.

A bacterium has one large circular loop of chromosomal material, as well as several much smaller rings known as plasmids.

Plasmids are easily removed from bacteria. They are small enough to be removed, genetically altered and put back into a bacterium.

The stages involved in the reprogramming of bacteria to produce a human protein, such as insulin, are as follows:

1   The section of DNA in the human cell that has the required gene, is identified on its original chromosome.

2   The gene is cut out of the chromosome using an enzyme.

3   A plasmid is removed from a bacterium.

4   The plasmid is cut open using the same enzyme.

5   The required gene is inserted into the plasmid using another type of enzyme.

6   This is repeated many times.

7   The altered plasmids are then inserted into bacteria.

8   The reprogrammed bacteria are given the correct conditions to reproduce, making many identical copies.

The bacteria that have been genetically modified (GM bacteria) with the transferred gene make large quantities of the gene's protein, if given suitable conditions.

One of the important advantages of using bacteria in this way is the rapid rate at which they reproduce. This means that the mass production of the desired protein using this method is a relatively quick process.

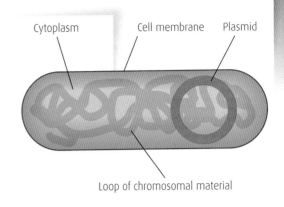

Cytoplasm   Cell membrane   Plasmid

Loop of chromosomal material

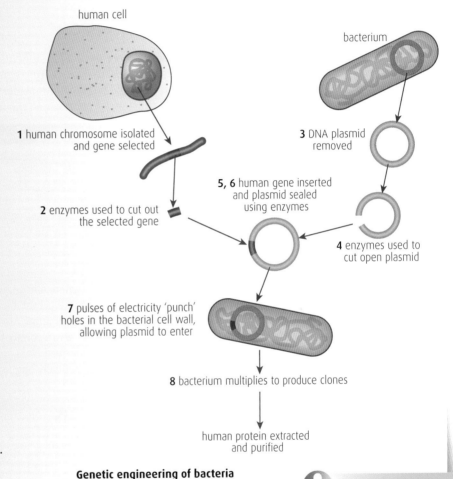

human cell

1 human chromosome isolated and gene selected

2 enzymes used to cut out the selected gene

bacterium

3 DNA plasmid removed

5, 6 human gene inserted and plasmid sealed using enzymes

4 enzymes used to cut open plasmid

7 pulses of electricity 'punch' holes in the bacterial cell wall, allowing plasmid to enter

8 bacterium multiplies to produce clones

human protein extracted and purified

Genetic engineering of bacteria

THINGS TO DO AND THINK ABOUT

Enzymes are used in several stages in the process of genetic engineering. Can you explain why the same enzyme can be used to cut the required gene from the human chromosome and to cut open the plasmid, but a different enzyme is required to put the gene into the plasmid?

DON'T FORGET

Genetic information can be transferred from one cell to another by genetic engineering.

ONLINE TEST

Test yourself on the transfer of genetic information online at www.brightredbooks.net/N5Biology

THE USES OF GENETIC ENGINEERING

Genetic engineering is growing in importance in our everyday lives. As new techniques are developed, more and more uses for genetic engineering are being discovered.

Genetic engineering has applications in medicine, as well as in commercial operations such as the food and drinks industry and in the manufacture of everyday materials like washing powders.

The following information shows applications of genetic engineering, but the details of each do not need to be learned.

ONLINE

Read more about the history and development of insulin, from early versions made using animal proteins to the genetically engineered version we use today, at www.brightredbooks.net/N5Biology

MEDICAL APPLICATIONS

As people are living longer, more medical treatments are needed than ever before. Genetic engineering is used now to manufacture a variety of products. Genes are extracted from human DNA and inserted into bacteria or yeast, where they produce human forms of useful substances, as shown in the table below.

Product of genetic engineering	Medical application
Insulin	Used to treat people who cannot make sufficient insulin naturally due to type 1 diabetes
Antibiotics	Used to treat people with bacterial infections – antibiotics prevent the growth of these microorganisms
Human growth hormone	Used to treat children who cannot make enough growth hormone naturally – human growth hormone prevents reduced growth
Factor VIII	Used to treat people with haemophilia who cannot make Factor VIII (and whose blood is, therefore, unable to clot)

There are several advantages to using genetically engineered products, like those given above, for medicinal purposes. One of these is that it is possible to produce a pure form of the protein which is free from viral contamination. Also, as the products are made using human genes, many people are ethically more comfortable with them. Previously, they had objections to using products that were derived from animals. Another benefit is that fewer people have allergic reactions to the genetically engineered products than to other forms.

COMMERCIAL APPLICATIONS

Several industries now rely on genetic engineering to manufacture their products. Some examples are given here.

Crop production

- Tomatoes were one of the first food crops to be genetically engineered. Genes were inserted which gave the tomatoes enhanced flavour. Other genes were introduced which gave the tomatoes a much longer shelf-life without the quality of the fruit deteriorating.

- GM potatoes have been genetically modified to be resistant to the disease called potato blight. This disease can destroy whole fields of potatoes, spoiling the entire crop. It is, therefore, of great advantage to grow potatoes with resistance to this disease.

- Rapeseed is an increasingly important food crop, as the seeds are used to make rapeseed oil. In recent years, researchers started developing new rapeseed varieties with the advantage of rapid growth. Original varieties contained a bitter-tasting chemical, making the oil unusable in food production. A toxic chemical was also present which prevented the plant being used as animal feed. The new varieties produced by genetic engineering lack both of these undesirable qualities and have led to the rise in the growth of the crop.

contd

- Genetic engineering has been used to create transgenic crops (plants with genes from another organism), including new strains of corn and soybeans. These crops contain genes which allow them to be resistant to certain herbicides. This means that farmers can spray their crops with herbicide to kill the weeds without damaging their crop plants. The crop field is kept free from weeds, increasing the crop yield. Resistance to herbicides has now been developed in a greater number of crop plants.

- β-carotene is a plant pigment which is important to humans because it is converted to vitamin A in our bodies. Rice plants produce β-carotene in their green tissues, but not in the edible part of the seed. Other valuable nutrients (for example, vitamin B and nutritious fats) are found in the outer coat of the rice grains, but are lost in the milling and polishing processes of white rice production. Unprocessed brown rice retains some of these nutrients, but is not suitable for long-term storage as it goes rancid. As rice is such an important part of the world population's diet, scientists wanted to improve its nutritional quality and turned to genetic engineering. Even though all the required genes to produce β-carotene are present in the grain, some of them are turned off during development. Scientists figured out how to turn on this complex pathway. Golden rice grains are easily recognisable by their yellow to orange colour. The stronger the colour, the more β-carotene is present.

White rice and golden rice

Other commercial applications

- UK scientists have created genetically modified chickens that do not spread bird flu. Bird flu is a disease caused by a virus and it spreads rapidly when poultry is produced in intensive conditions. An artificial gene has been inserted into chickens; this introduces a tiny part of the bird flu virus into the chickens' cells. They can still become infected but render the virus harmless to other poultry. Scientists believe that this genetic modification is harmless to the chickens and to people who might eat the birds. The researchers say that, in principle, the technique could be used to protect any farm animal from any disease. The eventual aim is to develop animals that are completely resistant to viral diseases.

- Making cheese involves curdling of milk. This was traditionally brought about by the use of an enzyme called rennin, which was extracted from the lining of calves' stomachs. Genetic engineers have now isolated the gene from the stomach lining and have transferred it to yeast cells. These yeast cells are grown in fermenters to produce large quantities of rennin, which is used by the cheese-making industry.

- Biological washing powders contain enzymes such as proteases to digest many of the stains that are found on clothing. These enzymes are produced by genetically reprogrammed bacteria. They are grown in large fermenters and the enzymes are extracted from them and added to the washing powders. These types of powders allow stains to be digested at lower temperatures, using less energy and saving money.

DON'T FORGET

When a particular desirable characteristic is identified, it is only the gene(s) for that characteristic that is transferred between organisms and not whole chromosomes.

ONLINE TEST

Test yourself on the uses of genetic information online at www.brightredbooks.net/N5Biology

THINGS TO DO AND THINK ABOUT

There is much debate about the use of genetically modified crops. Many countries have banned them, while others are happy to grow them. Can you find out why?

THE PURPOSE AND IMPORTANCE OF RESPIRATION

ENERGY FROM FOOD

Energy that is locked up in food is released in all living cells to allow chemical processes to take place. Green plants are able to make their own food through the process of photosynthesis, but animals must consume food to obtain energy.

Human foods contain three main components: fats (and oils), proteins and carbohydrates. These contain large molecules that must be broken down into smaller basic units to be absorbed into the body. The breakdown of large food molecules is called digestion. The table gives the basic units of the three main food components and the chemical elements they contain.

Food component	Basic unit
Carbohydrate	Glucose molecules
Fat/oil	Fatty acids and glycerol
Protein	Amino acids

A simple experiment can be carried out to investigate the energy value of each of the food types.

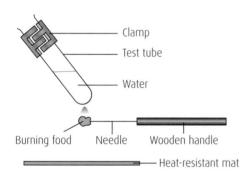

Clamp

Test tube

Water

Burning food   Needle   Wooden handle

Heat-resistant mat

Apparatus to measure the heat energy from a variety of food types

A measured volume of water is placed in the test tube and the temperature noted. A 1 g portion of a food type is then set alight and the flame used to heat the water in the tube until the food is completely burned. The temperature of the water is recorded and the rise in temperature calculated.

By repeating this experiment with 1 g of each of the three food types, a comparison can be drawn.

However, this method has several drawbacks:

● Not all of the heat energy from the burning food goes into the water. Some of it heats the test tube and the surrounding air.

● Not all of the food is burned. It is very difficult to make sure that the food stays alight until every last bit of energy is given out.

● It is very difficult to keep environmental factors from interfering with each experiment. For example, draughts in the room can blow the flame about.

For these reasons, a more sophisticated piece of apparatus called a calorimeter is used. This gives much more accurate energy values.

Using this method all of the heat energy released by the food is measured.

Results obtained from calorimeter experiments have shown that for equal quantities used, fat (or oil) contains about twice as much energy as protein or carbohydrate. This is why our bodies store energy as fat – more energy can be stored in a small quantity of material.

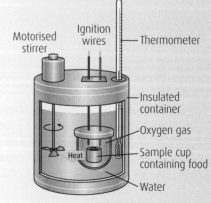

Motorised stirrer   Ignition wires   Thermometer

Insulated container

Oxygen gas

Heat

Sample cup containing food

Water

Food calorimeter

THE PROCESS OF RESPIRATION

To release the energy from food, living cells of both plants and animals must carry out a process known as **respiration**. The molecules which are broken down to release energy are known as respiratory substrates. The main respiratory substrate is glucose (a type of carbohydrate), but fats and proteins can be broken down if glucose is not available.

The carbohydrate, glucose, is a type of sugar. It consists of the chemical elements carbon (C), hydrogen (H) and oxygen (O); it can be broken down into these elements and reassembled into waste products, giving out energy in the process.

The overall reaction of respiration can be summarised by the following word equation:

glucose + oxygen → carbon dioxide + water + energy

In this reaction :

1   the glucose and the oxygen are the raw materials

2   the carbon dioxide and the water are the waste products

3   the purpose of the reaction is to release energy.

In the presence of oxygen, as shown by the equation above, glucose is broken down fully to release all of its energy and give the waste products, carbon dioxide and water. This is known as **aerobic respiration**.

Under certain circumstances, however, respiration can take place in the absence of oxygen and is then known as fermentation (see page 26). Different waste products are formed in this type of respiration.

ONLINE TEST

Test yourself on this topic by taking the 'Purpose and Importance of Respiration' test online at www.brightredbooks.net/N5Biology

THE ROLE OF ATP

Adenosine triphosphate (ATP) is a chemical found in all living cells. It consists of a molecule of adenosine with three inorganic phosphate (Pi) groups attached in a chain-like fashion.

ATP acts as a type of energy management system in cells. Just as money is placed into a bank to keep and is withdrawn when it is required, energy can be 'banked' in molecules of ATP and released when needed.

Structure of ATP

The energy released from ATP can be used in many cellular activities including muscle cell contraction, cell division, protein synthesis and transmission of nerve impulses.

THINGS TO DO AND THINK ABOUT

Explain why burning food in a calorimeter is a better method to use when investigating the energy value of food, compared to burning it on the end of a needle to heat water in a test tube.

DON'T FORGET

In aerobic respiration, the raw materials are glucose and oxygen, the waste products are carbon dioxide and water and the purpose of the reaction is to release energy.

THE CHEMISTRY OF RESPIRATION

The chemical energy stored in glucose is released by all cells through the series of enzyme-controlled reactions known as respiration.

Respiration takes place in several stages. The first stage always occurs in the cytoplasm, but further stages are dependent on whether or not oxygen is available.

VIDEO LINK

Check out the 'Aerobic Respiration' clip at www.brightredbooks.net/N5Biology

AEROBIC RESPIRATION

Aerobic respiration is the term given to the type of respiration which involves the breakdown of glucose in the presence of oxygen. Carbon dioxide and water are the products of the process.

$$\text{glucose} + \text{oxygen} \longrightarrow \text{carbon dioxide} + \text{water} + \text{energy}$$

The process of aerobic respiration starts when a molecule of glucose is broken down into two molecules of **pyruvate** in the cytoplasm. This happens as a series of reactions, each of which is controlled by enzymes. This process is known as **glycolysis**. 'Glyco' refers to carbohydrate (glucose) and 'lysis' means splitting. Therefore, the name indicates that a carbohydrate (glucose) is being split.

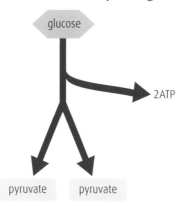

Splitting glucose

During this series of reactions, enough energy is released to synthesise two molecules of ATP.

Like glucose, pyruvate also contains a lot of energy locked up in the molecule. Each molecule of pyruvate moves into a mitochondrion and, here, a further series of enzyme-controlled reactions gradually breaks down the pyruvate to release this energy. During these reactions, carbon dioxide and water are formed and a lot more molecules of ATP are synthesised for every pyruvate that is broken down. The carbon dioxide and water are released as the waste products of respiration.

The complete set of respiration reactions is quite complex and a simplified version is shown below.

The reactions in the mitochondria allow the synthesis of a lot of molecules of ATP. Add to this the two ATP molecules which were formed as glucose was split in the cytoplasm and it can be seen that one molecule of glucose yields a large number of molecules of ATP in aerobic respiration.

DON'T FORGET

Aerobic respiration only takes place if oxygen is available.

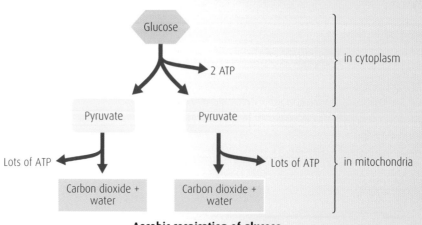

Aerobic respiration of glucose

RESPIRATION WITHOUT OXYGEN

A different type of respiration takes place if oxygen is not available.

The process begins in the cytoplasm in exactly the same way as in aerobic respiration: one molecule of glucose is broken down into two molecules of pyruvate and two molecules of ATP are synthesised. However, due to the lack of oxygen, the pyruvate molecules remain in the cytoplasm and the mitochondria do not become involved.

The pyruvate molecules are partially broken down by different chemical pathways depending on the type of organism carrying out the respiration.

Respiration without oxygen in multicellular animals

This type of respiration can occur in human muscle cells when a very heavy demand for more energy occurs and the oxygen supply is insufficient to meet the demand, for example when sprinting very fast.

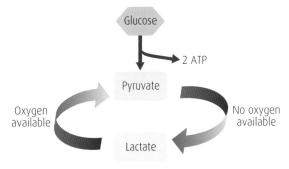

Fermentation in animal cells

In this case, pyruvate is only partly broken down, forming lactate, in a process called fermentation. No more ATP molecules are formed during this reaction. As the concentration of **lactate** builds up in the muscle cells, it causes muscle fatigue and pain. Since the muscles become inefficient, this process can only be maintained for a short period of time.

In order to get rid of the lactate, oxygen is required. When oxygen becomes available, the lactate is converted back to pyruvate, which can then enter the mitochondria for aerobic respiration to take place.

Since the lactate can be converted back into pyruvate, fermentation in animal cells is described as a reversible process. It is not a very efficient process, however, as no ATP is produced during the breakdown of pyruvate. So, one molecule of glucose yields only the 2 molecules of ATP formed initally, in fermentation in animal cells.

Respiration without oxygen in plants and microbes

This type of respiration can occur in plant cells that are deprived of oxygen, for example in waterlogged root cells, and also in microbes, such as yeast cells, which are important in the brewing industry.

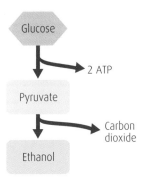

This involves the breakdown of pyruvate to ethanol (a type of alcohol) and the release of carbon dioxide gas. Again, no more ATP is formed in this reaction.

As carbon dioxide is released in the breakdown of pyruvate, fermentation in plants and microbes is an irreversible reaction. Again, it is not a very efficient process as no ATP is produced during the breakdown of pyruvate. So, one molecule of glucose yields only the 2 molecules of ATP formed initially, in fermentation in plants and microbes.

Fermentation in plant cells and microbes

COMPARISON OF RESPIRATION WITH AND WITHOUT OXYGEN

Respiration with oxygen	Respiration without oxygen
Glucose completely broken down	Glucose partly broken down
Carbon dioxide and water produced	Animals: lactate produced Plants and microbes: ethanol and carbon dioxide produced
A large number of molecules of ATP produced	Only 2 molecules of ATP produced

 ONLINE TEST

Test yourself on this topic by taking the 'Chemistry of Respiration' test online at www.brightredbooks.net/N5Biology

 THINGS TO DO AND THINK ABOUT

Explain why respiration with oxygen can be described as an efficient process, while fermentation is inefficient.

PRODUCING NEW CELLS

MITOSIS

All new cells are produced from existing cells by the process of cell division. Cell division occurs in single-celled organisms in order for them to reproduce. Each new cell is a complete organism.

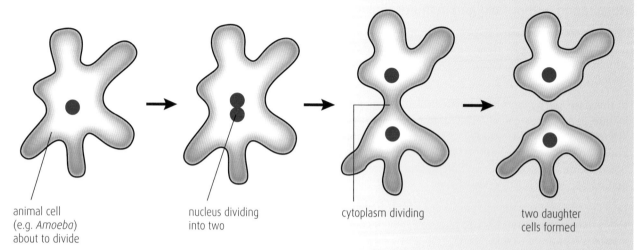

animal cell
(e.g. *Amoeba*)
about to divide

nucleus dividing
into two

cytoplasm dividing

two daughter
cells formed

DON'T FORGET

In mature animals which have stopped growing, the main purpose of cell division is replacement of worn out cells and repair of damaged tissue. When plants are mature, they continue to grow at root and shoot tips, as well as having to produce cells for repair.

In a multicellular organism, cell division is the process which increases the number of cells. This provides the means for growth and also for repair of damaged tissues.

An important part of cell division is the division of the cell nucleus. This is called **mitosis**. Mitosis ensures that each of the two daughter cells produced from the parent cell contains the same number of chromosomes as the parent cell and are genetically identical to it.

DON'T FORGET

Chromosome numbers can vary from organism to organism: a modern horse has 64 chromosomes, a duck has 80 and a cabbage has 18. However, often the number of chromosomes is the same in different organisms, although it is clear to see the differences between them. For example, a human and a privet plant each have 46 chromosomes, while a gorilla and a potato each have 48 chromosomes. It is the genetic make-up of the chromosomes that makes the individual, not the number of them.

MAINTAINING THE DIPLOID NUMBER IN MITOSIS

The nucleus contains thread-like structures known as chromosomes. They carry information in the form of **genes** which determine the characteristics of an organism. Normally, the individual chromosomes in a cell cannot be seen, but as the cell begins to divide, the chromosomes untangle, becoming shorter and thicker. This makes them visible using a microscope and they become free to move around the cell.

In multicellular organisms all cells, apart from the sex cells, have two sets of matching chromosomes. Human cells have two sets each consisting of 23 chromosomes in each set; that is, they have 23 pairs. Each chromosome in a pair is an identical match to the other in terms of the genes it carries.

The number of chromosomes in this 'double' set is known as the **diploid number**. The diploid number in humans is 46. When a cell divides, it is essential that the diploid number is maintained so the new cells receive a full set of genetic information. This is because the genetic information controls cell development and all of the activities that cells carry out. It is therefore vital that none of the information is missing.

THE SEQUENCE OF EVENTS IN MITOSIS

The events that take place in the process of mitosis ensure each daughter cell has the diploid number of chromosomes.

The following diagrams show the sequence of events.

Sequence of events in mitosis

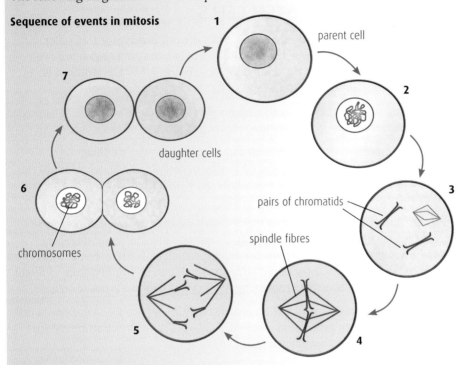

1 Each of the chromosomes, although not visible at this stage, makes a duplicate copy of itself (chromosomes are now pairs of **chromatids**).

2 The pairs of chromatids become shorter and thicker, and start to untangle.

3 The nuclear membrane disintegrates and the **spindle fibres** appear.

4 Each pair of chromatids attaches to one of the spindle fibres, at the **equator** of the cell.

5 The spindle fibres contract, separating the chromatids and pulling them to opposite ends of the cell.

6 A nuclear membrane begins to form around each group of chromosomes and the cytoplasm begins to divide.

7 Two new daughter cells form as the cell membrane divides the cells. Each new cell has an identical chromosome complement to the parent cell.

The sequence of events in mitosis is the same in both plant and animal cells. The main difference is found at the very end when the two newly formed cells separate from each other:

- In animal cells, the cell membrane pinches inwards from both sides until it meets in the middle, dividing the cytoplasm evenly.

- In plant cells, a new cell wall is formed as well, which divides the cell in two.

 ONLINE TEST

Check how well you've learned about mitosis at www.brightredbooks.net/N5Biology

 ## THINGS TO DO AND THINK ABOUT

Mitosis ensures that the diploid chromosome complement is maintained in the production of new cells. Explain why this is necessary.

STEM CELLS 1

AN INTRODUCTION TO STEM CELLS

As every cell in a multicellular organism arises from the original zygote formed at **fertilisation**, it follows that every cell has the full set of genetic information required to construct the entire organism. However, only some of the genes are 'switched on' in each cell, giving rise to differentiation or specialisation for a particular role.

It is this switching on and off of genes that allows the development of the great variety of specialised cells that exists in the tissues, organs and systems of an organism.

Most cells in an organism become specialised at a fairly early stage in their development. Once a cell has become altered to suit its function, it cannot revert back to its original form or develop into any other type of cell and is known as a permanent tissue cell.

Stem cells are different from permanent tissue cells. One of the main features of stem cells is that they have the ability to multiply, (self-renew), while still maintaining the potential to develop into other cell types. For example, stem cells can become cells of the heart, blood, brain, bones, muscle and skin. Stem cells can be sourced from different areas of the body, but all have this capacity to develop into multiple types of cells.

Stem cells are involved in the growth of an organism. They can produce any type of cell that is required. Even when growth has stopped, stem cells have an important role to play in replacing worn out and damaged cells and tissues.

Stem cells

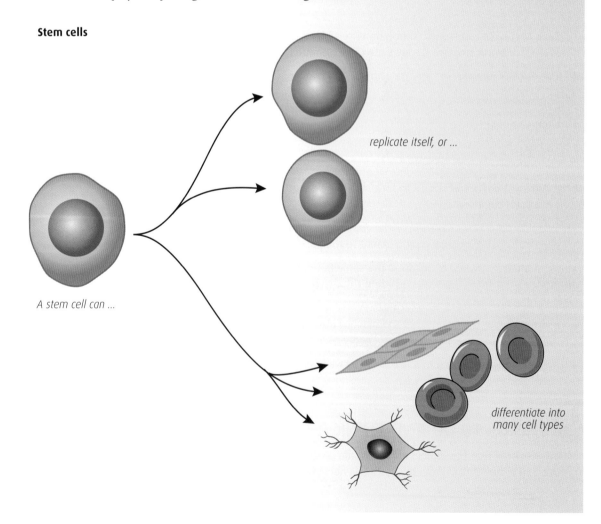

replicate itself, or ...

A stem cell can ...

differentiate into many cell types

TYPES OF STEM CELLS

- Embryonic stem cells are present in embryos. Embryonic stem cells form during the very early stages of development, when the zygote is dividing rapidly to form a ball of cells. These cells retain the capacity to divide for long periods of time and have the ability to make all of the various cell types required in the organism.

- Adult stem cells are obtained from some of the tissues in the body of an adult. Bone marrow, which is found in the long bones of a human, is a rich source of stem cells and it is used in the treatment of some types of cancer and blood diseases like leukaemia. Research by scientists has led to the development and isolation of stem cells from adult human skin cells and this work is still ongoing.

THE DISCOVERY OF STEM CELLS

The potential of stem cells to develop into any type of cell was recognised in the 1980s, but it was towards the very end of the 20th century that scientists found methods of isolating these cells from human embryos and growing them in the laboratory.

Most of the work was done using cells from embryos which had been created in the laboratory for the treatment of infertility. They were embryos which were 'left over' after the treatment and were no longer required for that purpose. This required consent from the individuals whose embryos were donated to this type of research.

A further development was the identification of conditions that allow some specialised adult human cells to be genetically reprogrammed to assume an embryonic stem cell-like state.

By using tissue culture techniques, stem cells can be induced to grow and divide in a controlled way, producing a collection of continually dividing undifferentiated cells. This collection is known as a stem cell line and has many potential uses. Once established, a cell line can be grown indefinitely in the laboratory and cells may be frozen for storage.

VIDEO LINK

For a video about stem cells, watch 'A stem cell story' at www.brightredbooks.net/N5Biology

ONLINE TEST

Test yourself on this topic by taking the 'Stem cells 1' test online at www.brightredbooks.net/N5Biology

DON'T FORGET

Adult stem cells can be found in several areas of the body such as in the blood, skin and brain. However, they are not as versatile as embryonic stem cells. Embryonic stem cells can develop into any type of cell, but in adults blood stem cells, for example, can only develop into the different types of blood cell.

THINGS TO DO AND THINK ABOUT

1. In what way are stem cells different from the cells which have become permanent tissue?

2. Why are stem cells involved in growth and repair of tissue?

STEM CELLS 2

THE USES OF STEM CELLS

Research and treatments

Stem cells are used in research to increase understanding of cell development. They are also important in the treatment of some conditions such as cancer and some birth defects.

Replacement of damaged tissues

Damaged or diseased organs are often replaced with transplants of healthy organs from a donor. However, the demand for such organs and tissues is far greater than the available supply. Research is taking place into the possibility of growing stem cells which are then directed to develop into specific types of cells, providing a source of new cells, tissues and even organs for transplant. This would be a renewable resource as stem cells can be grown continuously, given correct conditions. It is hoped that stem cells could be involved in the treatment of:

- Brain diseases – such as Parkinson's disease, in which damage to nervous tissues causes muscles to move continually, and Alzheimer's disease, in which damage to brain cells results in confusion and memory loss. Embryonic cells have been 'programmed' to develop into brain cells to replace damaged ones.

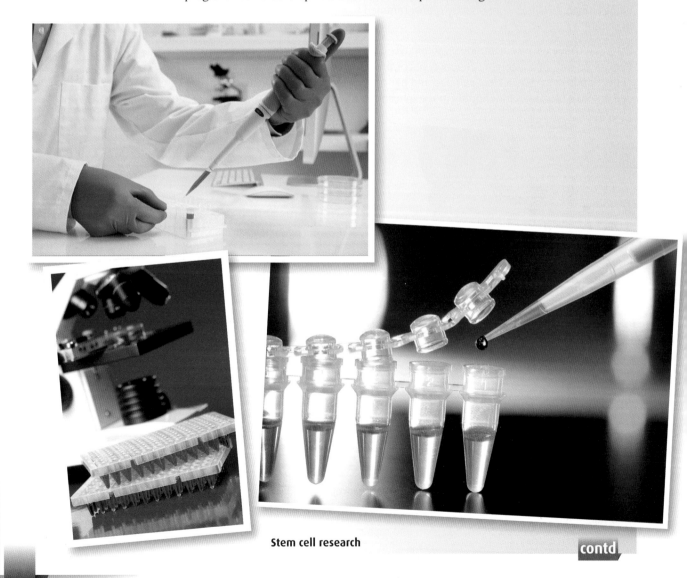

Stem cell research

contd

- Blood diseases – such as leukaemia, in which white blood cells grow and divide out of control, and sickle cell anaemia, in which red blood cells are misshapen and have reduced oxygen carrying capacity. As there are difficulties associated with bone marrow stem-cell extractions, treatment has been developed using the stem cells found in the umbilical cord and the placenta of newborn babies. This has led some scientists to suggest that umbilical cord blood should be stored for future use.

- Heart disease – this is another area which is being explored for treatment by stem cells. For example, the possibility of growing heart muscle cells from stem cells in the laboratory is being investigated. These healthy muscle cells could be transplanted into patients who suffer from chronic heart disease.

- Type 1 **diabetes** – this results when the cells of the **pancreas** which normally produce **insulin** are destroyed by the person's own immune system. Investigations suggest that embryonic stem cells could be cultured to form insulin-producing cells which might then be transplanted into people with diabetes.

Testing of new drugs

Stem cells could be grown in the laboratory and used to test drugs and chemicals before they are trialled in people. This means that drugs could be tested on human cells, rather than on animals, resulting in safer drugs that are developed in a more ethical way.

Screening for toxins

Stem cells could be used to grow tissues for testing our response to certain chemicals. For example, stem cells could be used to investigate the effects of pesticides which are sprayed onto food crops. The pesticides could be applied to the stem cells to see if there were any adverse effects.

These are only some of the ways in which stem cells are being used or may be used in the future, and it is an area of continued significant research and development.

ETHICAL ISSUES

Stem cells have been used in medicine for many years, for example in bone marrow transplants.

Ethical issues about stem cells focus almost entirely on embryonic stem cells, and use of these cells does pose a dilemma. It forces a choice between two moral principles:

1   the sense of duty involved in preventing or relieving suffering

2   the sense of duty to respect the value of human life.

The basic problem is that, in using embryonic stem cells, the embryo is destroyed. This means the loss of potential human life. However, this has to be weighed up against the fact that stem cell research could lead to the discovery of new medicines and treatments that could alleviate suffering and prolong life.

Much debate has taken place at the highest court levels in many countries and this is still an area which is highly controversial.

THINGS TO DO AND THINK ABOUT

1   Where are stem cells found in humans?

2   Why are stem cells important?

3   Why might some people be against the use of stem cells for medical treatment while others are in favour of it?

DON'T FORGET

Stem cells differ from other body cells in that they retain the ability to divide and develop into any type of specialised cell.

ONLINE

Check out the US Department of Health and Information's guide 'Stem cell information' at www.brightredbooks.net/N5Biology

ONLINE TEST

Test yourself on this topic by taking the 'Stem cells 2' test online at www.brightredbooks.net/N5Biology

CELLS, TISSUES AND ORGANS: THE ORGANISATION OF CELLS AND MULTICELLULAR ORGANISMS 1

The basic unit of life is a cell. It is the smallest independent life form. Some cells exist as individual organisms and other cells are a part of more complex multicellular organisms.

UNICELLULAR ORGANISMS

Some organisms consist of only one cell, which must show all the characteristics of life and carry out all of the functions required for survival. These organisms are called unicellular organisms. Examples are bacteria, some algae, some fungi (including yeast) and a group of organisms called protozoa.

Unicellular organisms are unfamiliar to most of us as they are generally too small to be seen with the naked eye. Most protozoa are around 0·01–0·05 mm, although some can grow as large as 0·5 mm. They can, however, be easily seen using a microscope.

A few examples of unicellular organisms are shown here.

DON'T FORGET

Unicellular organisms are often measured in micrometres (μm) because of their tiny size. Therefore, a cell of 0·01 mm would be 10 μm.

Unicellular organisms

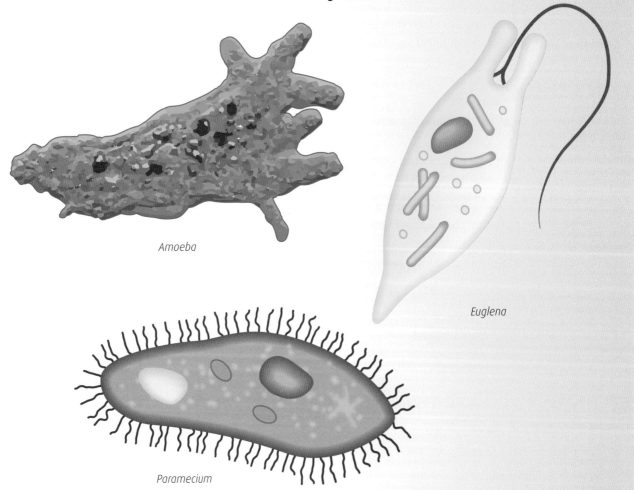

Amoeba

Euglena

Paramecium

contd

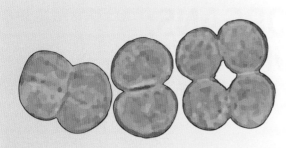

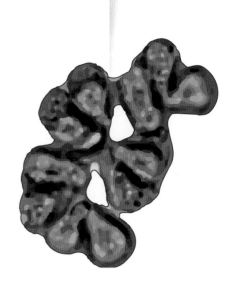

Pleurococcus

Single-celled algae

Diatom

As shown in earlier chapters, some unicellular organisms are extremely useful to humans, for example bacteria are used in genetic engineering and yeast cells are used in fermentation to produce alcohol. However, some are disease-causing cells and are known as pathogens.

MULTICELLULAR ORGANISMS

More familiar to us are organisms which are made of many cells. These are known as multicellular organisms. In multicellular organisms, it would be inefficient for every cell to carry out all the processes required for survival. Instead, a division of labour exists, where cells are specialised so that each type of cell carries out a particular function.

Each cell of a multicellular organism has a dual role to play. It must be independent and able to act at an individual level, as well as working together with neighbouring cells to perform a part in the operation of the whole organism.

At the individual level, cells must be able to absorb food, absorb oxygen, carry out respiration and manufacture proteins, for example. These are basic-level survival functions. In addition, however, the same cells must work together with groups of similar cells to perform a specialised function that benefits the entire organism, aiding its survival.

THINGS TO DO AND THINK ABOUT

1　Name one unicellular organism that is plant-like and one that is animal-like.

2　What type of organism is yeast?

3　Living organisms show seven different characteristics of life. Can you name them?

ONLINE

For more on this, check out the 'Unicellular v multicellular' link at www.brightredbooks.net/N5Biology

ONLINE TEST

Test yourself on this topic by taking the 'The organisation of cells and multicellular organisms 1' test online at www.brightredbooks.net/N5Biology

DON'T FORGET

In unicellular organisms, the cell must be able to carry out all of the functions of living things. In multicellular organisms, there is a division of labour, with each cell fulfilling a specific function.

CELLS, TISSUES AND ORGANS: THE ORGANISATION OF CELLS AND MULTICELLULAR ORGANISMS 2

CELL ORGANISATION

Groups of cells that are similar in structure and work together to carry out a similar function are known as **tissues**.

The diagram shows some of the tissues in the knee of the human body.

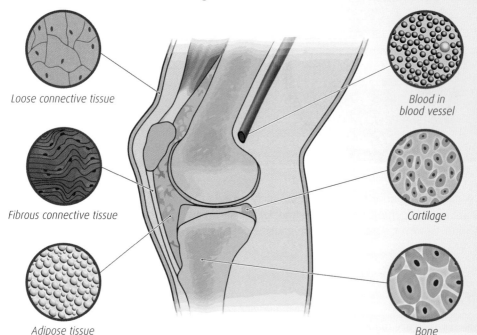

Loose connective tissue

Fibrous connective tissue

Adipose tissue

Blood in blood vessel

Cartilage

Bone

A structure such as the human knee contains a variety of different tissues

From the diagram, we can see that not all cells in a tissue have to be exactly the same, although this can be the case. For example, the cells which make up **cartilage** tissue can all be one type. In contrast to this are the cells which make up **blood**. Here different cell types (**red blood cells** and **white blood cells**) are present. These cells vary in shape, content, size and individual function, but still work together, as part of the tissue.

Plants also have a level of organisation in which cells operate together as tissues.

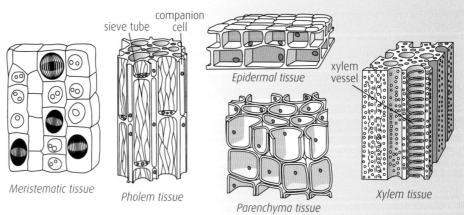

companion cell

sieve tube

Meristematic tissue

Pholem tissue

Epidermal tissue

xylem vessel

Parenchyma tissue

Xylem tissue

Plant tissues

As with animals, plant tissues can be made up of groups of cells that are all of a similar type, such as meristematic cells, or can consist of a variety of cell types working together. An example of this is **phloem** tissue, which consists of cells which form sieve tubes and cells called **companion cells**. These cell types differ in structure and content, but together deliver the overall function of transport of food materials in the plant.

contd

In order for efficient functioning of multicellular organisms, tissues are organised into organs, which are part of systems. Systems are coordinated together to perform as a whole functioning organism.

Thus, there is an organisational hierarchy of cells in a complex multicellular organism.

Cells ➡ Tissues ➡ Organs ➡ Systems ➡ Organism

Hierarchy of organisation of cells

The stomach is an **organ** found in the human digestive system. It is made up of a variety of tissues which consist of many types of cells. The stomach is just one part of the digestive system.

An organ is a highly specialised structure in which a collection of tissues is joined in a structural unit to serve a common function. Some organs may have more than one function.

Examples of human organs are **heart**, **liver**, brain and lungs.

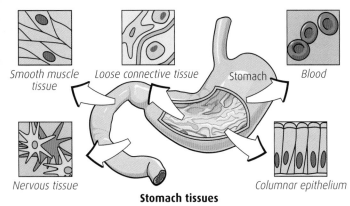

Smooth muscle tissue Loose connective tissue Stomach Blood

Nervous tissue **Stomach tissues** Columnar epithelium

- The heart acts as a pump and consists of tissues such as muscle, nerve and blood.

- The liver breaks down poisonous chemicals and produces bile to aid digestion. It consists of connective tissue, blood and nerves.

- The brain processes information and examples of tissues found here include nervous tissue, blood and blood vessels.

- The lungs are involved in gas exchange and contain epithelial tissue, as well as blood.

Plants show similar organisation. Tissues work together to make organs which are part of systems.

Examples of plant organs are stem, leaves and roots.

- The stem supports the leaves of the plant. A variety of tissues such as **xylem** and phloem are found here.

- The leaves are the main photosynthetic organs of the plant and contain tissues such as **epidermis** and palisade **mesophyll**.

- The roots anchor the plant, and absorb water and minerals through xylem tissue and receive food through phloem tissue.

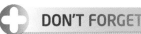

stem phloem xylem

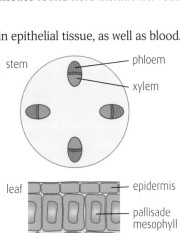

leaf epidermis pallisade mesophyll

 THINGS TO DO AND THINK ABOUT

1 Name two types of cells found in animal blood tissue.

2 Name two types of tissues found in a plant's transport system and for each tissue, state a cell type found there.

3 Why is it important that cells are specialised in a multicellular organism?

4 Starting with individual cells, describe the hierarchical organisation of cells in a living organism.

CELLS, TISSUES AND ORGANS: SPECIALISED CELLS AND THEIR FUNCTIONS 1

Multicellular organisms are made up of many different cell types, which are organised into tissues and organs. So, when cells are produced by cell division, they must undergo changes to become specialised for their particular function. Genes control the development of newly formed cells and bring about the many changes that produce the great variety of cells found in a complex multicellular organism.

The basic idea that a cell becomes modified to suit a particular function is important, although it is not necessary to know the individual details of how this happens. You don't need to memorise all of the examples which follow but you do need to have an awareness that cells differentiate to become specialised and you should able to give an example of this taking place.

SPECIALISATION OF ANIMAL CELLS

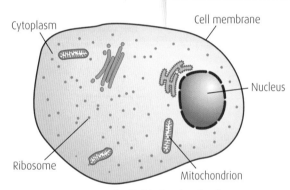

A newly divided animal cell

Cytoplasm
Cell membrane
Nucleus
Ribosome
Mitochondrion

When an animal cell is produced, it has the same basic structure as all other newly produced animal cells. As it develops, its contents and shape are altered to make it specialised. Exactly what happens to each basic cell depends on where it is found in the organism, which organ it is associated with and the function it has to carry out.

A simplified version of a newly divided animal cell is shown in the diagram.

Let's look at some examples of how cells like this are modified to become specialised for a particular function.

Motor neuron

As can be seen from the diagram, many changes have taken place to make a motor **neuron** from an unspecialised cell. The function of this type of cell is to transmit electrical impulses from the **central nervous system** to the muscles or glands of the body, causing a response to some sort of **stimulus**. The cell might be very long in length as it may carry information from, for example, the spinal cord to the foot. Some of the specialised adaptations to form a motor neuron are:

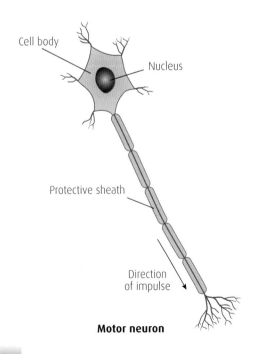

Cell body
Nucleus
Protective sheath
Direction of impulse

Motor neuron

- the main part of the cell has become elongated to form a conducting fibre

- extensions have been created at the ends so that connections can be made with other cells

- a protective sheath has formed to ensure that the electrical impulses travel from one end of the neuron to the other.

contd

Sperm cell

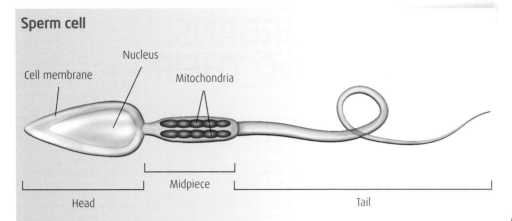

Human sperm cell

Again, many changes have taken place to create a sperm cell from the original unspecialised cell. The function of this type of cell is to carry half the genetic code to the egg and to fertilise it, creating a single cell called a **zygote**. This can go on to develop into an embryo. Some of the adaptations that take place to form a sperm cell are:

- the cell has become elongated and has three distinct areas
- the 'head' section has become small with the DNA tightly condensed in the nucleus
- many mitochondria have been produced and are found in the 'midpiece'
- the 'tail' section has developed to allow the sperm to swim
- organelles that are not required have disintegrated.

Red blood cell

The function of a red blood cell is to transport oxygen around the body. It is part of blood tissue, which forms part of the circulatory system. Red blood cells contain the pigment **haemoglobin**, which combines with oxygen to form **oxyhaemoglobin**. The specialisation required to form a red blood cell includes:

- the nucleus of the cell has disintegrated, leaving more space in the cell for the transport of oxygen
- haemoglobin has been produced
- the cell shape has been modified to become biconcave (curving inwards on both sides) to increase its surface area, allowing faster absorption of oxygen.

Red blood cells

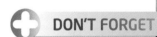

ONLINE

Check out the 'Cell specialisation' link at www.brightredbooks.net/N5Biology

DON'T FORGET

Cells must become specialised so that they can take on a specific role in the efficient functioning of the whole multicellular organism.

ONLINE TEST

Test yourself on this topic by taking the 'Specialised cells and their functions 1' test online at www.brightredbooks.net/N5Biology

THINGS TO DO AND THINK ABOUT

1   You should be able to name a tissue and give an example of how the cells have become specialised to allow them to carry out their particular function.

2   Why does a red blood cell have no nucleus? Describe one other feature of this cell which allows it to carry out its role efficiently.

CELLS, TISSUES AND ORGANS: SPECIALISED CELLS AND THEIR FUNCTIONS 2

SPECIALISATION OF PLANT CELLS

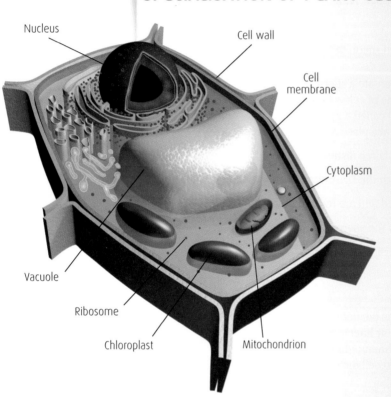

Nucleus

Cell wall

Cell membrane

Cytoplasm

Vacuole

Ribosome

Chloroplast

Mitochondrion

A newly divided plant cell

When a plant cell is produced, it has the same basic structure as all other newly produced plant cells until it undergoes the changes that make it specialised. As with animals, the exact detail of what happens to the basic cell depends on where it is found, which organ it is associated with and the function it has to carry out.

Plant cells are produced in certain areas of the plant known as **meristems**. A simplified version of a newly divided plant cell is shown in the diagram.

Although the diagram shows the presence of chloroplasts in the cell, not all plant cells possess these. This depends on whether the cell is positioned in an area of the plant which receives light and on whether it is involved in the process of photosynthesis.

Let's have a look at some examples of how plant cells become modified for a particular function.

A sieve tube cell

A sieve tube cell forms part of the phloem tissue. The function of the sieve tube cell is to transport sugar made in the leaves by photosynthesis. It allows sugar to be provided to areas that are unable to carry out photosynthesis, such as the roots. It also transports sugar to storage organs both above and below ground. Changes that happen to make a sieve tube cell specialised for this function include:

- cells have become elongated

- end walls have become perforated, forming sieve plates

- cell organelles which are not required have disintegrated.

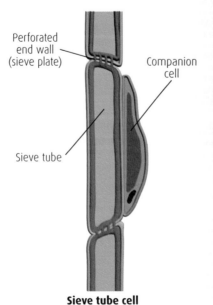

Perforated end wall (sieve plate)

Companion cell

Sieve tube

Sieve tube cell

A palisade mesophyll cell

The function of this type of cell is primarily photosynthesis. These are column shaped cells found near the upper surface of the leaf. They are closely packed together, forming a continuous layer of cells that are specialised for the absorption of light for photosynthesis. Specialisation of these cells involves:

- cells have become column shaped

- a large quantity of chloroplasts have been made in the cell to provide more chlorophyll for maximum absorption of light.

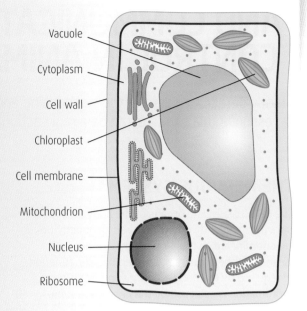

Vacuole

Cytoplasm

Cell wall

Chloroplast

Cell membrane

Mitochondrion

Nucleus

Ribosome

Palisade mesophyll cell

A ROOT HAIR CELL

This type of cell is, again, a modification of the basic unspecialised plant cell. Its function is to absorb water from the soil into the root of the plant. The plant cell has been specialised to make it more efficient at this task and the modifications include:

- no chloroplasts have been produced (this cell does not receive light, so cannot photosynthesise)

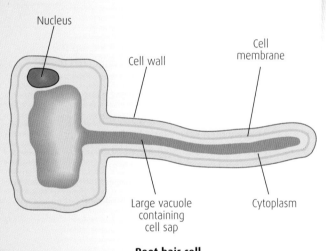

Nucleus

Cell wall

Cell membrane

Large vacuole containing cell sap

Cytoplasm

Root hair cell

- the cell has grown an 'extension', known as the **root hair**, to increase the surface area for water absorption

- the vacuole has extended into the root hair to maximise the osmotic difference between the cell and the surrounding cell water, to make the osmosis of water and the diffusion of dissolved minerals more efficient.

ONLINE

Check out the 'Root hair cells and osmosis' link at www.brightredbooks.net/ N5Biology

DON'T FORGET

Although there are some organelles that are present in all living cells, the function of the cell determines the actual organelles that it needs.

ONLINE TEST

Test yourself on this topic by taking the 'Specialised cells and their functions 2' test online at www. brightredbooks.net/ N5Biology

THINGS TO DO AND THINK ABOUT

1   You should be able to describe the way the cells in a tissue become specialised to allow them to carry out their particular function.

2   Describe the way in which a typical plant cell becomes specialised to carry out the function of absorbing water from the soil.

3   Why does a palisade mesophyll cell have more chloroplasts than other cells in a leaf?

CONTROL AND COMMUNICATION: NERVOUS CONTROL IN ANIMALS

The body needs to constantly detect stimuli from the surrounding environment and make appropriate responses. A response to a stimulus can be a rapid action from a muscle or a slower response from a gland.

THE NERVOUS SYSTEM

The cells, tissues and organs of the body do not work independently of one another; their activities are coordinated so that they work together. This means that they can perform their specific function as required by the body at any given time. This coordination is achieved by the nervous system.

The human nervous system is composed of three parts:

- the brain
- the spinal cord
- the nerves.

The brain and the spinal cord together make up the central nervous system (CNS). This is connected to all parts of the body through the nerves, which are made up of thousands of long thin nerve fibres.

In order to protect itself, the body must be able to detect changes which take place both internally and externally. These changes act as stimuli. Nerves lead to and from all of the organs and systems of the body, making sure that information is gathered and appropriate responses are made.

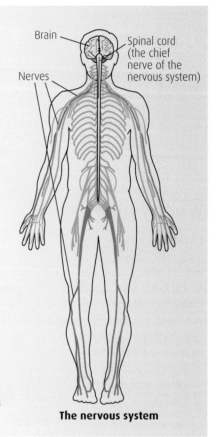

The nervous system

NEURONS

The nervous system is made up of nerve cells called neurons. These cells are specialised to carry electrical impulses. These impulses are only able to travel in one direction along the neuron. There are neurons that send impulses to the CNS, different neurons within the CNS to process impulses and other neurons which send impulses to organs which carry out responses.

There are, therefore, three basic types of neuron and each has a different function as summarised in the table.

Type of neuron	Function
Sensory neuron	To send information from the receptors of the sense organs to the CNS
Inter neuron	To send information within the CNS
Motor neuron	To send information from the CNS to the muscles and glands

The flow of information follows a pathway as shown in the diagram.

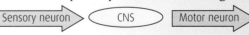

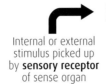

Internal or external stimulus picked up by **sensory receptor** of sense organ

Response by muscle or gland (effector)

Flow of information in the CNS

AREAS OF THE BRAIN

The brain is made up of millions of neurons which are coordinated to process information. There are three main areas of the brain: the **cerebrum**, the **cerebellum** and the **medulla**.

The cerebrum

This is the largest part of the brain and is divided into two halves or hemispheres. This part of the brain is responsible for receiving information from the sensory neurons, processing it and sending impulses to all parts of the body along the motor neurons in order to make some sort of response. It processes information from the senses such as sight and hearing. The cerebrum is also involved in mental and conscious thought processes such as reasoning, imagination, memory and thinking.

The cerebellum

This area of the brain is found towards the rear of the head. Its primary function is to control balance and coordinate muscle actions. Complex actions which require skill in both these areas, such as riding a bicycle, involve the cerebellum.

The medulla

The medulla is found at the top of the spinal cord. It controls involuntary body processes such as breathing, digestion, heart rate, swallowing and sneezing.

Communication within the brain is by short neurons which pass electrical impulses very quickly. Coming from the main body of the cell are lots of tiny extensions which connect with other cells and receive information. This information (in the form of electrical impulses) is sent along a thin fibre which has a layer of insulation, preventing loss of the impulse as it travels to the other end of the neuron. Here, chemical transmitters transfer the information to the next cell.

Neurons are highly specialised cells and, because they carry electrical impulses, they do not actually come into contact with one another. This means that there has to be a way of transferring the information across a gap between two neurons. This is carried out by chemical messengers called neurotransmitters in the following process:

1   An electrical impulse travels from the cell body of a neuron along the nerve fibre to the other end.

2   When it arrives, this triggers the nerve ending of that neuron to release chemical messengers called neurotransmitters.

3   The neurotransmitters diffuse across the gap between the first neuron and the next neuron. The gap is known as a **synapse**.

4   The neurotransmitters bind with receptor molecules on the membrane of the second neuron. The receptor molecules are specific to the chemicals released from the first neuron.

5   The second neuron now becomes stimulated to transmit the electrical impulse.

Synapses occur between all types of neurons and are necessary because the neurons do not touch each other directly.

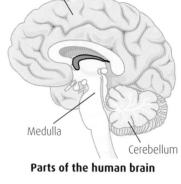

Cerebrum

Medulla

Cerebellum

Parts of the human brain

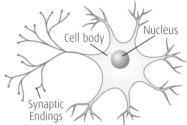

Cell body

Nucleus

Synaptic Endings

Neuron in the brain

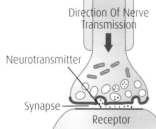

Direction Of Nerve Transmission

Neurotransmitter

Synapse

Receptor

A synapse

 ONLINE

For more on neurons, check out the 'Nervous system' link at www.brightredbooks.net/N5Biology

 DON'T FORGET

The brain and the spinal cord make up the CNS (central nervous system). The brain is composed of many different areas and each area is responsible for a different function. The three main areas are the cerebrum, cerebellum and medulla.

 ONLINE

For more information, check out 'How does your brain work?' at www.brightredbooks.net/N5Biology

 ONLINE TEST

Test yourself on this topic by taking the 'Control and communication 1' test online at www.brightredbooks.net/N5Biology

 THINGS TO DO AND THINK ABOUT

A synapse is found at the junction between two nerve cells. Describe what happens at a synapse.

CONTROL AND COMMUNICATION: NEURONS AND REFLEX ACTIONS

Nerves and the CNS provide communication within the body. Sensory neurons gather information about the internal and the external environment, pass the information in the form of electrical impulses to the CNS and then impulses travel along motor neurons to effectors (muscles and glands) to make appropriate responses.

TYPES OF NEURON

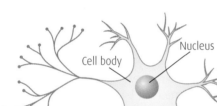

Sensory neuron

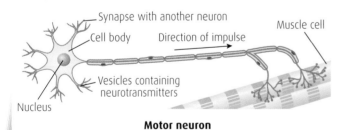

Inter neuron

Cell body
Nucleus
Receptors to detect stimuli
Direction of impulse – towards CNS

Nucleus
Cell body
Synaptic Endings

Synapse with another neuron
Cell body
Direction of impulse
Muscle cell
Vesicles containing neurotransmitters
Nucleus

Motor neuron

Sensory neuron

The function of a sensory neuron is to send an electrical impulse to the CNS in response to a stimulus from a tissue or organ of the body. This could be an internal stimulus, such as the swallowing of food or an external one, such as the eye (as a sense organ) seeing a ball flying through the air.

A sensory neuron has extensions at the sensory end of the nerve fibre, where receptors detect sensory input or stimuli. This allows the gathering of information. The main body of the cell is part way along the length of the cell fibre and at the other end of the sensory neuron are more extensions, ready to send neurotransmitters across the synapse to the next cell.

Inter neuron

The function of an inter neuron is to send an electrical impulse within the CNS inside the spinal cord or within the brain. Inter neurons pass information from sensory neurons to other neurons. They are generally shorter in size than the other neurons.

Motor neuron

The function of the motor neuron is to send an electrical impulse from the CNS to the muscles or glands, which then make a response. This could be to produce an enzyme, such as pepsin in the stomach, or to cause muscles to contract in the arms and hands to catch a ball.

A motor neuron has extensions at the end of the nerve fibre and these are connected to the muscle or gland, allowing the motor neuron to pass on an electrical impulse which will bring about a response. The main body of the cell is at one end of the fibre and has receptors to receive neurotransmitters which cross the synapses from neighbouring cells.

These three types of neurons work together, receiving stimuli, processing the information and making appropriate responses for the safe functioning of the organism.

VIDEO LINK

Check out the clip 'Structure functions and types of neurons' at www.brightredbooks.net/N5Biology

REFLEX ACTIONS

A **reflex action** is a type of response to a stimulus which does not need to be learned and which does not involve conscious thought. For example, if a person puts their hand down on something which is very hot, their response does not require thought or learning. The hand is automatically pulled away; it is an involuntary response. This action is known as a reflex action and it protects the body from harm.

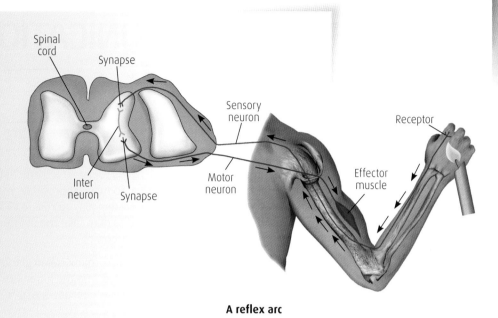

A reflex arc

The pathway that brings about a reflex action is called a **reflex arc** and is illustrated in the diagram.

The steps in the reflex arc are:

1 The stimulus is detected by receptors in the sense organ, the skin, (in the example above, pain receptors in the hand pick up the stimulus).

2 An electrical impulse is immediately sent along the sensory neuron towards the spinal cord.

3 The impulse crosses the first synapse and travels along the inter neuron.

4 The impulse crosses the second synapse and is picked up by the motor neuron.

5 The motor neuron conducts the impulse to the nerve endings in the muscle.

6 The muscle contracts, removing the hand from the stimulus which was causing the pain.

Reflex actions protect the body from damage. Other examples of reflex actions are:

- blinking when an object travels towards the eye
- the changing size of the eye's pupil in response to varying light levels
- swallowing when food reaches the back of the throat
- a jerking reaction when the knee is tapped.

Reflex actions often take place via the spinal cord and without involving the brain; this cuts down on the time that it takes to process the information, allowing a very fast response.

 ONLINE TEST

Test yourself on this topic by taking the 'Control and communication 2' test online at www.brightredbooks.net/N5Biology

 DON'T FORGET

A reflex action is a rapid, involuntary response to a potentially harmful stimulus.

 ## THINGS TO DO AND THINK ABOUT

1 Give a definition of a reflex action and explain why these actions are important.

2 Describe the pathway of information sent through a reflex arc from the moment a person stands on a sharp object, such as a tack, to the appropriate response being made.

3 Which of the following are examples of reflex actions?
   a) Sneezing when dust enters the nose.
   b) Kicking a ball in a game of football.
   c) Running to avoid an oncoming car.
   d) Blinking when a ball comes towards your head.
   e) The pupil of your eye constricting in bright light.

CONTROL AND COMMUNICATION: HORMONAL CONTROL

THE ENDOCRINE SYSTEM

In order to protect itself, the body needs to detect information about conditions – both internal and external – and then to make appropriate responses. The preceding chapters deal with how the nervous system plays a role in this function. Another system, the **endocrine system**, also has a role to play. Like the nervous system, this system is involved in communication around the body.

The endocrine system consists of several organs called glands. These glands release chemical substances, known as **hormones**, into the bloodstream. Hormones are chemical messengers. They carry information from one area of the body to another.

HORMONES VERSUS NERVE IMPULSES

Both hormones and nerve impulses work to coordinate the activities of the body, but they act in very different ways. When a nerve impulse is triggered by a stimulus, an electrical impulse travels along a particular route (the nerve fibre) to arrive at a specific muscle or gland. In contrast, when an **endocrine gland** is stimulated to release a hormone, the hormone enters the bloodstream and travels around the whole body. This means that the hormone is in the blood that arrives at every tissue, although they will not all react to it. So, to summarise, nerve impulses only travel along one particular route, while hormones travel around the entire body.

Consider this example. Imagine making a phone call to one of your friends. The information you tell that person goes straight to them and only them. That is similar to the way in which a nerve impulse travels – to one specified place. In contrast, however, if you were to go on television and deliver the same message, it would be broadcast to millions of people – but only those concerned about the message would respond. Similarly, hormones travel in the blood to all parts of the body, but only some parts respond.

Many hormones are proteins. They are secreted (released) following a specific stimulus. Although hormones are sent to all areas of the body, most of the tissues they arrive at 'ignore' their presence. It is only when a hormone arrives at a tissue which has cells with the correct hormone receptor (the target tissue), that the hormone has an effect.

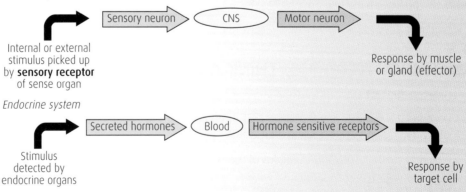

Nervous system

Internal or external stimulus picked up by **sensory receptor** of sense organ → Sensory neuron → CNS → Motor neuron → Response by muscle or gland (effector)

Endocrine system

Stimulus detected by endocrine organs → Secreted hormones → Blood → Hormone sensitive receptors → Response by target cell

Comparing nervous and endocrine systems

HORMONE RECEPTORS

Hormone receptors are found embedded in the surface membrane of certain cells. A target tissue has cells with complementary receptor proteins for specific hormones, so only that tissue will be affected by these hormones. When a hormone arrives at a cell with the correct receptor, having diffused out of the bloodstream, it attaches to the cell membrane. Each hormone only attaches to a matching receptor. In the diagram, the cell from tissue 1 has the necessary receptor sites for the hormone (and, therefore, is sensitive to it), but the cell from tissue 2 lacks the required receptor and does not react to the presence of the hormone.

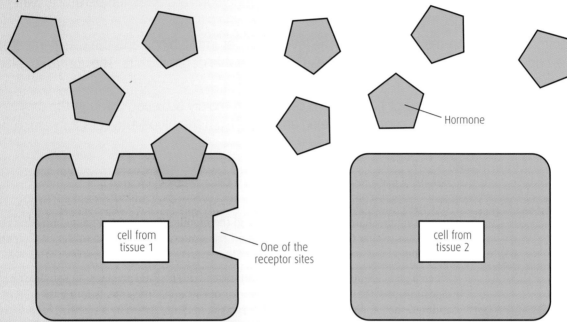

Hormone and target cell

Once a hormone has combined with the receptor on the cell membrane, a response occurs in the cytoplasm of that cell.

Although hormones are produced in tiny quantities, they have very important effects. The effects of some types of hormones can be long lasting. These are able to control the size to which the body grows, and can have an effect on mental development and personality. Other types of hormone have a more immediate effect and cause changes which last for a shorter period of time.

THINGS TO DO AND THINK ABOUT

1   Give two similarities and two differences between the control of body functions by the nervous system and the control by the endocrine system.

2   Place these statements in the correct order to show the sequence of events in hormonal control:

hormone binds to target cell receptors

stimulus is detected by endocrine organ

target cell makes a response

hormone is released

hormone travels in the bloodstream

3   Which of the control mechanisms, hormones or nerve impulses, is capable of having the longer lasting effect?

ONLINE TEST

Test yourself on this topic by taking the 'Control and communication 3' test online at www.brightredbooks.net/N5Biology

ONLINE

For more, check out the 'Endocrine system: facts, functions and diseases' at www.brightredbooks.net/N5Biology

DON'T FORGET

The nervous system relies on nerve impulses to carry information from one place to another in the body. The endocrine system operates by sending chemical messengers (hormones) in the bloodstream to travel to all parts of the body.

CONTROL AND COMMUNICATION: THE CONTROL OF BLOOD GLUCOSE LEVEL

It is essential that the conditions inside the body remain at the optimum for efficient function. There are systems in place that constantly monitor these conditions and that make sure that they are kept within tolerable boundaries. Some important conditions are body temperature, water content and blood sugar level.

A monitoring system in the brain detects the level of water in the blood and the core temperature of the body, and it makes adjustments if these are not within defined limits. The control of blood sugar is brought about by the action of the pancreas.

The body requires the release of energy in all cells. This is mainly achieved through the process of aerobic respiration. Glucose is continually being absorbed from the blood into cells for respiration.

MONITORING GLUCOSE LEVELS

To ensure that there is a constant supply of glucose for cell respiration, the body has regulating mechanisms. Glucose levels in the bloodstream are carefully controlled to make sure that there is sufficient sugar present in the blood for respiration to take place, but not so much that it stops the other systems of the body working properly.

When a meal is eaten, carbohydrates are digested. The final product of carbohydrate digestion, glucose, passes from the interior of the small intestine, through the villi walls and into the capillaries of the circulatory system. This causes a rise in the blood sugar level. The body has to be able to cope with the fact that we eat relatively large quantities of food at a time (raising the blood sugar level considerably) and then it can be quite some time before we eat again (causing the blood sugar level to drop substantially). The body monitors rising and falling glucose level and adjusts it accordingly.

THE PANCREAS

The pancreas is an organ situated just behind the stomach.

As well as producing digestive enzymes, the pancreas has a role to play in the regulation of blood sugar level. It is able to do this as it is an endocrine organ which produces two hormones, **insulin** and **glucagon**. These hormones are released directly into the bloodstream and have an immediate effect on blood sugar:

- insulin lowers the blood sugar level.

- glucagon raises the blood sugar level.

These two hormones work together to make sure that blood glucose is maintained at the correct level.

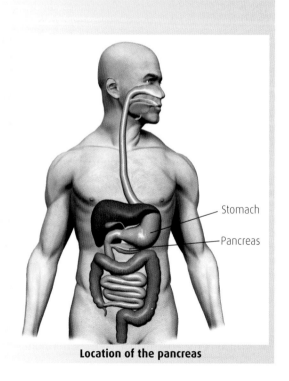

Stomach

Pancreas

Location of the pancreas

VIDEO LINK

For more on how the pancreas works, check out the clip 'Endocrine system, pancreas' at www.brightredbooks.net/N5Biology

THE LIVER

Another organ involved in the regulation of blood glucose is the liver. It is the largest internal organ in the body and is located just below the diaphragm, to the right of the stomach. It acts as a kind of storage facility for carbohydrate, removing glucose from the blood when the level is high and releasing glucose back into the blood when the level drops.

A rise in blood sugar to above the normal level is detected by receptor cells in the pancreas as blood flows through this organ. These receptor cells respond by producing the hormone insulin. The insulin is carried in the bloodstream to the liver. Here, insulin stimulates the liver cells to produce an enzyme that converts glucose to **glycogen** (a storage carbohydrate). About 100g of glycogen can be stored in the liver at any one time.

Glycogen is insoluble, making it a good storage compound. Because it is insoluble it has no osmotic effect. This is in contrast to glucose which, being soluble, can affect the balance of water in tissues.

If a long time has passed since a meal was eaten, the blood sugar level might drop below the normal level. This, again, is detected by a part of the pancreas as the blood flows through. A different response is triggered. The production of insulin slows down and a different hormone called glucagon is produced. This travels in the blood to the liver where it activates another enzyme that converts stored glycogen back to glucose. This is released into the blood from the liver, increasing the blood sugar level and returning it to the normal level.

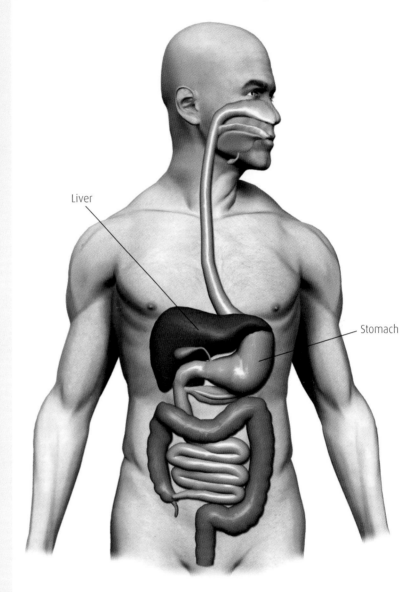

Location of the liver

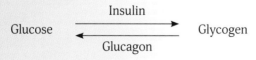

$$\text{Glucose} \xrightleftharpoons[\text{Glucagon}]{\text{Insulin}} \text{Glycogen}$$

THINGS TO DO AND THINK ABOUT

Name the organ which is responsible for monitoring the blood sugar level and state its location in the body.

ONLINE TEST

Test yourself on this topic by taking the 'Control and communication 4' test online at www.brightredbooks.net/N5Biology

DON'T FORGET

Insulin lowers the blood sugar level by converting glucose to glycogen for storage. Glucagon raises the blood sugar level by converting glycogen back to glucose.

CONTROL AND COMMUNICATION: DIABETES

The disease diabetes is caused by the inability of some or all of the cells in the pancreas to make insulin. This lack of insulin causes the concentration of glucose in the blood to rise and remain above the normal level.

There are two main types of diabetes, known simply as type 1 and type 2.

ONLINE

Check out the NHS guide 'Diabetes, type 1' for more information at www.brightredbooks.net/N5Biology

TYPE 1 DIABETES

This type of diabetes is the form that is most common in children, but it can develop at any age. It is an autoimmune disease, meaning that the body's defence system mistakenly attacks certain cells in the pancreas, permanently destroying them. These cells can no longer produce insulin.

Without insulin, glucose cannot be moved out of the blood into the cells and the concentration of glucose in the blood increases. This can seriously damage the organs of the body. The body cells start to break down fat and muscle, leading to weight loss. Other symptoms include being very thirsty, needing to urinate often, feeling excessively tired and having skin infections.

The cause of type 1 diabetes is likely to be a combination of genetics and an environmental trigger. It is not yet fully understood why the body's immune system attacks the pancreas.

People with type 1 diabetes need to inject insulin regularly to manage their diabetes. It is a life-long condition and cannot be reversed. A carefully calculated diet, planned physical activity and regular blood glucose testing are required. Diabetes is a serious condition and it can lead to complications such as kidney disease, heart disease and stroke if left untreated.

TYPE 2 DIABETES

Type 2 diabetes occurs for one of two reasons: the cells of the pancreas fail to produce enough insulin to keep the blood sugar level normal or the cells of the body no longer react to the insulin that is produced, meaning that they cannot take up the glucose which is in the blood. This is known as insulin resistance.

Type 2 diabetes is a much more common disorder that type 1. It can potentially be avoided through correct exercise and diet. If a person develops extreme insulin resistance, they may need to take tablets or inject insulin to keep their blood sugar level stable.

This type of diabetes was once called adult-onset diabetes, but it is now becoming more common in children and teenagers. This increase has been linked to the increase in obesity. Other factors which can increase the risk of developing type 2 diabetes include having high blood pressure, having high cholesterol, being overweight and having a close family member with type 2 diabetes. The chances also increase with age.

Facts about type 2 diabetes:

- Type 2 diabetes is estimated to affect over 2·5 million people in the UK.
- If either parent has type 2 diabetes, the risk of inheriting it is 15%.
- If both parents have type 2 diabetes, the risk of inheriting it is 75%.
- Almost 1 in 3 people with type 2 diabetes develop the complication of kidney disease.

THINGS TO DO AND THINK ABOUT

1. Explain the roles of insulin and glucagon in maintaining a normal blood sugar level.
2. Describe what can go wrong in the body to cause diabetes.

DON'T FORGET

Diabetes is caused by a blockage in the hormone communication pathway due to a failure in the release of insulin or a failure to respond normally to insulin.

VIDEO LINK

Check out the NHS guide 'Diabetes, type 2' for more information at www.brightredbooks.net/N5Biology

ONLINE TEST

Test yourself on this topic by taking the 'Control and communication 5' test online at www.brightredbooks.net/N5Biology

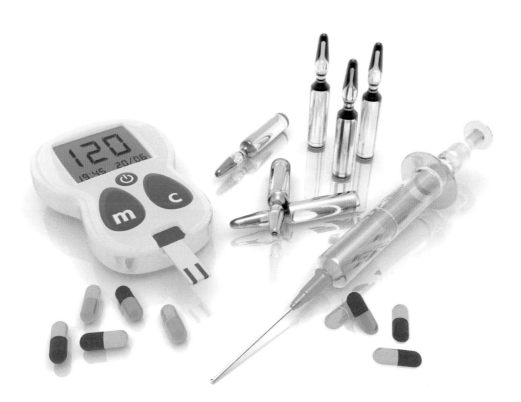

REPRODUCTION: THE STRUCTURES AND SITES OF GAMETE PRODUCTION

A **gamete** is a cell that fuses with another cell at fertilisation. Gamete is, therefore, another name for sex cell. Flowering plants produce gametes, as do animals that reproduce sexually.

GAMETE PRODUCTION IN PLANTS

Flowers are the organs of sexual reproduction in plants. Although the appearance of individual flowers can be quite different, the basic structures are the same. Most often, the flower contains both male and female parts. The diagram shows the main structures of a flower.

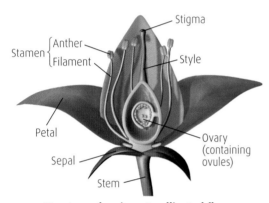

Structure of an insect-pollinated flower

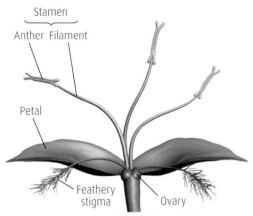

Structure of a wind-pollinated flower

Together, the filament and the anther make up the stamen. This is the male part of the flower and the anther produces pollen grains, which contain the male gametes.

The stigma, style and ovary make up the female part of the flower and the ovary produces ovules, which contain the female gametes.

The following table shows the function of each of the parts of the flower.

Part of flower	Function
Sepal	Protects the flower bud before opening
Petal	May be used to attract insects (often brightly coloured)
Stamen – anther	Produces pollen grains which each contain a male gamete
Stamen – filament	Holds the anther upright
Stigma	Traps pollen grains
Style	Leads from stigma to ovary
Ovary	Produces ovules
Ovule	Contains the female gamete

The type of flower shown in the top diagram has brightly coloured petals, may be pleasantly perfumed and has its sex organs (anthers and ovaries) inside the flower. These types of flowers are designed to attract insects for the process of pollination. Not all flowers have this structure. Some flowers are pollinated by the wind and, therefore, have no need for bright colours or strong scents. These flowers often have small greenish or brown petals and their sex organs hang out of the flower, exposed to the wind. The grass flower shown in the lower diagram is an example of a wind-pollinated flower.

POLLINATION

Once the flower has reached maturity and the pollen and ovules have been produced, the process of pollination takes place. Pollination is the transfer of pollen from the anther to the stigma. Usually, pollen from one flower lands on the stigma of another flower, but this can occur within the same flower. Insect-pollinated flowers have sticky stigmas, so that the pollen grains stay in place when they land there. Wind-pollinated flowers have stigmas that are feathery and hang out of the flower, acting like a net for trapping pollen grains.

Further development only takes place if the pollen lands on a stigma belonging to a flower of the same species.

FERTILISATION

After pollination has taken place, the pollen grain begins to grow a pollen tube. The tube grows through the tissues of the style and continues on towards the ovary. As it grows, the nucleus inside the pollen grain gradually makes its way down inside the tube. When the end of the pollen tube reaches an ovule in the ovary, it enters through a tiny hole. The tip of the pollen tube bursts, releasing the male gamete, which then fuses with the female gamete inside the ovule.

Fertilisation is the process by which the nucleus of the male gamete fuses with the nucleus of the female gamete to form a single cell. This is called a zygote. By repeated cell division, the zygote becomes a ball of cells which will eventually become an embryo plant. The ovule develops into a seed, acting as a food store for the embryo. The ovary, now containing the seeds, undergoes changes to become a fruit.

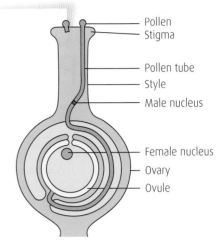

Growth of the pollen tube following pollination

GAMETE PRODUCTION IN ANIMALS

Like plants, animals produce gametes in specialised sex organs. In human males, sperm are produced in the testes and in females, eggs (or ova) are produced in the ovaries.

Sperm cells are the smallest human cells. They have a head containing a nucleus with a small amount of cytoplasm, some mitochondria and a long tail.

Millions of sperm cells are produced in the testes at any one time. They are able to swim as long as they are in a fluid, by moving their tails backwards and forwards. They are released from the testes and travel along the sperm duct, leaving the male body through the penis.

Egg cells are the largest human cells. An egg has a nucleus and much more cytoplasm than the sperm cell. Egg cells are unable to move by themselves and rely on other parts of the reproductive system to move them from one place to another.

Usually, only one egg cell is produced at a time. It is released from an ovary and cilia inside the oviduct help the egg to travel along towards the uterus. If the egg cell meets a sperm cell and fertilisation occurs in the oviduct, a zygote forms. This divides many times to form an embryo, which will then embed into the uterus wall to be nourished as it develops into a fetus.

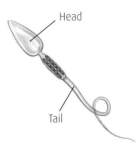

Male reproductive system

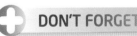

A sperm cell

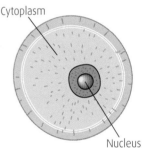

Egg cell

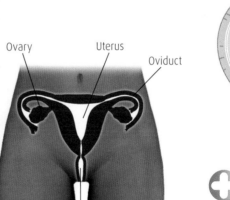

Female reproductive system

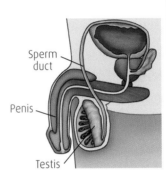

DON'T FORGET

'Gamete' is the name for any sex cell, whether it is produced by a plant or an animal.

THINGS TO DO AND THINK ABOUT

1   Name the sites of production of the male and female gametes in a flowering plant.

2   What is pollination?

3   Name the male and female gametes in humans.

4   Give a definition of the term 'fertilisation'.

ONLINE TEST

Test yourself on this topic by taking the 'The structures and sites of gamete production' test online at www.brightredbooks.net/N5Biology

REPRODUCTION: DIPLOID BODY CELLS AND HAPLOID GAMETES

SEXUAL REPRODUCTION AND VARIATION

Variation can be categorised as **continuous** or **discrete** (this is explored in the next chapter). Sexual reproduction gives rise to the variation which exists among members of a species. This is because the single cell formed at fertilisation (zygote) contains genes from each of the two parents and has, therefore, a new combination of genetic information.

The nucleus of every cell contains chromosomes. The number of chromosomes in cells varies from species to species, but is always the same in every body cell in a particular plant or animal. The number of chromosomes in each body cell is known as the **diploid** number. A diploid cell is one that contains two sets of chromosomes. One set of chromosomes comes from each parent.

The diploid number of a cell is represented as 2n, where n is the number of chromosomes in one set; so, 2n is a simple way of stating that the cell has two sets of chromosomes. In humans the 2n number is 46. This means that humans have two sets of 23 chromosomes.

A zygote forms when fertilisation takes place. The zygote is the result of two gametes joining together. Each gamete brings one whole set of chromosomes (the n number). When the gametes fuse, they create a zygote with two sets of chromosomes (the 2n number).

Organism	Diploid number (2n)	Haploid number (n)
Human	46	23
Chicken	78	39
Horse	64	32
Pineapple	50	25
Potato	48	24
Wheat	42	21
Lyon	38	19

While cells which contain two sets of chromosomes are described as diploid, cells such as gametes which have only one set of chromosomes are described as **haploid**. It is essential that gametes are haploid so that a zygote, produced by the joining of two gametes, can be diploid. As all body cells are produced from the repeated division of the zygote, all body cells have the diploid number of chromosomes.

Examples of diploid and haploid numbers of chromosomes of different organisms are shown in the table.

DON'T FORGET

The haploid number (n) is only found in the gametes; the diploid number (2n) is found in the nucleus of all other body cells.

HUMAN CELLS AS EXAMPLES OF ANIMAL CELLS

The male gamete, sperm, has one set of 23 chromosomes in its nucleus. The female gamete, the egg, also has one set of 23 chromosomes in its nucleus.

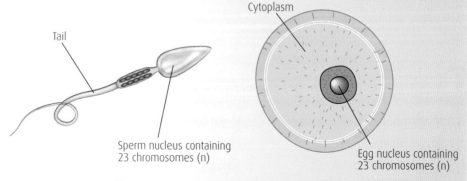

Tail

Sperm nucleus containing 23 chromosomes (n)

Cytoplasm

Egg nucleus containing 23 chromosomes (n)

Chromosome complement in human gametes

contd

When fertilisation takes place, both the male and female gamete nuclei join together to form a zygote containing the chromosomes from each of the gametes. The zygote, therefore, has two sets of 23 chromosomes, making a total of 46 altogether:

- 23 is the haploid number of chromosomes in a human

- 46 is the diploid number.

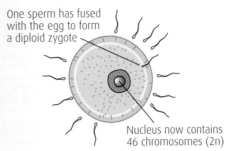

One sperm has fused with the egg to form a diploid zygote

Nucleus now contains 46 chromosomes (2n)

Diploid human zygote formed at fertilisation

ROSE CELLS AS EXAMPLES OF PLANT CELLS

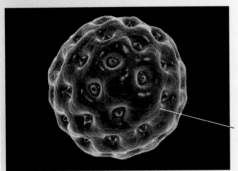

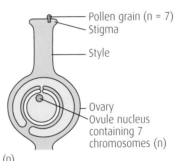

Pollen grain (n = 7)
Stigma
Style
Ovary
Ovule nucleus containing 7 chromosomes (n)
Pollen grain nucleus containing 7 chromosomes (n)

Chromosome complement in rose gametes

When fertilisation takes place, both nuclei join together to form a zygote containing the chromosomes from each of the gametes. The zygote therefore has two sets of seven chromosomes, making a total of 14 altogether:

- 7 is the haploid number of chromosomes in a rose

- 14 is the diploid number.

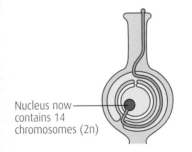

Nucleus now contains 14 chromosomes (2n)

Diploid rose zygote formed at fertilisation

THINGS TO DO AND THINK ABOUT

1 The table shows various types of cells. Decide whether each of them contains the haploid or the diploid number of cells for that species and tick the correct box.

2 Explain why all gametes are haploid.

3 Complete the table to give the haploid or diploid number for each of the organisms shown.

Type of cell	Haploid number	Diploid number
Cell from a human liver		
Pollen grain from a daffodil		
Skin cell from a rabbit		
Cheek cell from a pig		
Sperm cell from a salmon		
Ovule from a daisy		

Type of organism	Haploid number	Diploid number
Red deer		68
Cotton plant		52
Earthworm	18	
Cabbage		18
Cow	30	
Cat	19	

VARIATION AND INHERITANCE 1

Shape of aspen leaves

Pattern of cone shells

Human faces

AN INTRODUCTION TO VARIATION

The members of a species are not identical, even though they all possess genetic information for the same range of characteristics. Individuals show variations which make them different from one another.

Some variations may be due to effects of the environment which influence the development of an individual. These variations are unimportant to the species as a whole because they are not passed on from parent to offspring.

Other variations are caused by differences in the genetic information of individuals and these can be inherited. Sexual reproduction involves combining genetic information from both parents. This allows mixing of genes in different ways and so contributes to variation. The photographs show some examples of variation between members of the same species.

Genes can exist in different forms, each capable of producing a variant of a particular characteristic. The different forms of a **gene** are called **alleles**.

DISCRETE VARIATION

Discrete variation of a characteristic shows only a limited number of distinct possibilities. This type of variation is found in characteristics that are coded by a single gene with a limited number of forms or alleles.

Discrete variation has been important in the study of inheritance. Characteristics which have easily recognised variants are observed in successive generations. The patterns of their inheritance have allowed researchers to work out the mechanism involved.

Examples of discrete variation include:

- cat hair length

A) short hair

B) long hair

- tongue-rolling ability.

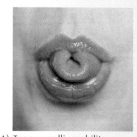

A) Tongue-rolling ability

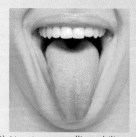

B) Non-tongue rolling ability

- blood groups

There are four possible blood groups. These are:

1   Group A   2   Group B   3   Group AB   4   Group O

CONTINUOUS VARIATION

Continuous variation of a characteristic shows a continuous range of possibilities between a minimum and a maximum value. There are no distinct groups and an individual's characteristic may have a value anywhere in the overall range of possibilities.

Continuous variation occurs because several different genes influence the same characteristic. Such a characteristic is said to be **polygenic**. When a number of genes contribute to a characteristic, it means that there are many different combinations of the various alleles involved. This produces many possible values for that characteristic, forming a continuous range of possibilities.

Examples of continuous variation include height, weight and hand span. Most visible characteristics are polygenic. It is probable that even some of those characteristics which show discrete variation and that are explained by single gene inheritance are influenced by more than one gene.

When the values for a polygenic characteristic are collected for a large number of individuals, it is found that they always show the same pattern of distribution. Relatively few individuals show values close to the extremes of the range. Most individuals show values close to the middle of the range, in other words a value close to the average.

This type of distribution is called a normal distribution. When it is plotted as a graph or chart, it shows a typical bell-shaped curve.

The photograph below shows a small group of people, all from the same university department, standing in order of their heights (in feet and inches).

The graph which follows it shows the distribution of their heights (in centimetres). The distribution is not a perfect normal distribution but the dotted line shows the overall pattern. If the sample size had been greater (more people included) then we would expect the pattern of height distribution to be closer to the typical normal distribution.

ONLINE TEST

Check how well you've learned about variation online at www. brightredbooks.net/ N5Biology

DON'T FORGET

The production of haploid gametes from diploid body cells and the random combination of gametes at fertilisation both contribute to genetic variation.

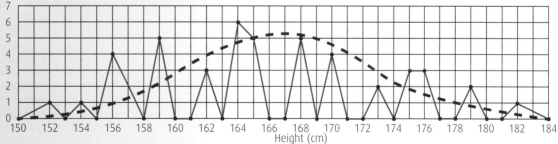

The distribution of heights in a group of people

THINGS TO DO AND THINK ABOUT

1. What is the most common height of the people in the group?

2. Can you calculate their average height to the nearest centimetre?

VARIATION AND INHERITANCE 2

INHERITANCE OF GENETIC INFORMATION

Inheritance is the passage of genetic information from parent to offspring. The characteristics that are coded in that information can be followed from one generation to the next. Organisms receive genetic information from both parents, giving them two complete sets of chromosomes (diploid). Therefore, they possess two pieces of genetic information (or two alleles) for every characteristic.

IMPORTANT DEFINITIONS FOR GENETICS AND INHERITANCE

Term	Definition
Gene	A unit of genetic information which codes for a protein. It controls a characteristic of an organism.
Allele	One of a number of alternative forms of a gene. Different alleles may code for different variations of a characteristic, for example tongue rolling and non-tongue rolling.
Phenotype	This is a description of a characteristic as it appears in an individual. For example; free ear lobes, green pea seeds and blood group A are all phenotypes.
Genotype	This is the inherited information of an organism. When studying inheritance the genotype usually refers to the information for only one or two characteristics, rather than all the genetic information of the organism. In other words the genotype identifies the particular alleles of a gene that the organism possesses.
Dominant	When two different alleles of a gene are present in an organism, the dominant allele shows its effect and the effect of the other allele is masked. For example, if a person carries both the allele for tongue rolling ability and the allele for non-tongue rolling, the person will be able to roll their tongue because the tongue rolling allele is dominant over the non-tongue rolling allele.
Recessive	This is the opposite of dominant. It refers to the allele which does not show its effect when two different alleles for a characteristic are present. For example, the allele for non-tongue rolling is recessive to the allele for tongue rolling.
Homozygous	If the two alleles that an organism possesses for a characteristic are the same, the organism is said to be homozygous for that characteristic. For example, if a person has two alleles for tongue rolling ability, they are homozygous for tongue rolling. If they have two alleles for non-tongue rolling, they are homozygous for non-tongue rolling.
Heterozygous	If the two alleles that an organism possesses for a characteristic are different, the organism is said to be heterozygous for that characteristic. For example, if a person has one allele for tongue rolling ability and one allele for non-tongue rolling, they are heterozygous for tongue rolling. They will be able to roll their tongue because that is the dominant trait.
P	The symbol used to represent the parental generation of a genetic cross.
F_1	The symbol used to represent the 1st generation of offspring of a genetic cross (1st filial generation).
F_2	The symbol used to represent the 2nd generation of offspring of a genetic cross (2nd filial generation).
Diploid	This refers to cells which contain two complete sets of chromosomes. These are derived from the one set inherited from each parent. The diploid number is represented as 2n, where n = the number of chromosomes in a single set. Normal body cells are diploid.
Haploid	This refers to cells which contain one complete set of chromosomes. The haploid number is represented as n, where n = the number of chromosomes in a single set. Sex cells, or gametes, are haploid.

contd

The diagram shows the inheritance of coat colour in several generations of pet mice.

The black coat colour allele is dominant to the allele for white coats.

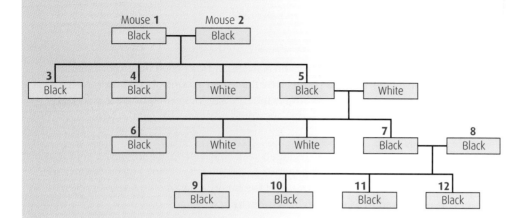

Alleles are often represented by an upper case letter for the dominant allele and the equivalent lower case letter for the recessive allele. In this example, **B** represents the allele for black coat colour and **b** represents the allele for white coat colour.

The phenotype for the coat colour of each mouse is shown. Using this information, the genotypes of some of the mice can be deduced.

All the white mice must have the **bb** genotype because the presence of just one **B** allele would give them black coats.

Mice 1, 2 and 5 must have the **Bb** genotype because they have each produced white offspring. A white mouse must receive a **b** allele from each parent.

Mice 6 and 7 must have **Bb** genotypes because they had a white parent and each must have received a **b** allele from it.

The genotypes of mice 3, 4, 8, 9, 10, 11 and 12 cannot be deduced from the available information. They may be **BB** or **Bb**. It is probable that at least one of mice G or H has the **BB** allele since all of their offspring are black, but this could be due to chance.

ONLINE

Learn more about this topic by watching the clip at www.brightredbooks.net

ONLINE TEST

Test yourself on variation and inheritance at www.brightredbooks.net

 THINGS TO DO AND THINK ABOUT

1 Explain how organisms carry two alleles for each characteristic.

2 Using the symbols **R** and **r**, give the three possible genotypes and their resulting phenotypes for tongue rolling in humans.

3 In the diagram showing the inheritance of coat colour in mice, how many mice make up the F_1 generation of mice 1 and 2?

VARIATION AND INHERITANCE 3

GREGOR MENDEL

Gregor Mendel was an Austrian monk and a scientist. He carried out experiments on inheritance using garden pea plants. He established groups of pea plants with a particular characteristic which was always inherited by their offspring. He referred to these plants as 'true-breeding'. Today, we would now refer to them as being homozygous.

The characteristics Mendel studied included seed shape, seed colour, flower colour, and pod colour. Each characteristic has only two possibilities and shows discrete variation. In many of his experiments, Mendel studied the inheritance of only one characteristic at a time. This is referred to as **monohybrid inheritance**.

Characteristic	Seed shape	Seed colour	Flower colour	Pod colour
Possible phenotypes	Round seeds	Yellow seeds	Purple flowers	Green pods
	Wrinkled seeds	Green seeds	White flower	Yellow pods

Mendel's peas: four of the seven characteristics studied by Mendel

MONOHYBRID CROSSES

One of Mendel's experiments was to cross homozygous round seeded plants with homozygous wrinkled seeded plants. These were his **P** generation.

All the offspring produced round seeds. These were his F_1 generation.

F_1 plants were crossed together to produce the F_2 generation. This contained both round seeded, wrinkled seeded and white-flowered plants in a 3:1 ratio.

Mendel carried out similar crosses using homozygous plants with the other pairs of contrasting phenotypes. His results were always the same. The F_1 generation always contained only one phenotype – the dominant one. The F_2 generation always contained both phenotypes in a 3:1 ratio of the dominant:recessive phenotypes.

Mendel explained these results by stating that organisms possess pairs of hereditary factors for each characteristic and that these factors separate during the formation of gametes (sex cells). Therefore offspring receive one factor (allele) from each parent at fertilisation.

contd

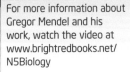

VIDEO LINK

For more information about Gregor Mendel and his work, watch the video at www.brightredbooks.net/N5Biology

THE MECHANISM OF INHERITANCE

For the cross described above

- **R** is used to represent the allele for round seed shape

- **r** is used to represent the allele for wrinkled seed shape.

P generation phenotypes	round-seeded plants	×	wrinkled-seeded plants
P generation genotypes	**RR**	×	**rr**
P generation gamete genotypes	all **R**		all **r**
F_1 generation genotype		all **Rr**	
F_1 generation phenotype		all round seeded	
F_1 generation genotypes	**Rr**	×	**Rr**
F_1 generation gamete genotypes	50% **R** and 50% **r**		50% **R** and 50% **r**

F_2 generation genotypes can be the result of a number of different possible combinations of gametes. These can be determined using a **Punnett square**.

		Gametes from first parent	
		R	**r**
Gametes from second parent	**R**	**RR**	**Rr**
	r	**Rr**	**rr**

F_2 genotypes	RR Rr Rr	rr
F_2 phenotypes	round seeded plants	wrinkled seeded plants
		(3:1 ratio)

Punnett squares can be used to predict the results of any cross where the genotypes of the parents are known.

It must be remembered that Punnet squares predict the expected phenotype ratios for any cross. There is no guarantee that these ratios will be achieved. This is because the fertilisations which take place between the gametes of the two parents are random.

The Punnett square above shows that an **R** allele from one of the parents has an equal chance of fusing with an **R** or an **r** allele from the other parent. This might happen but there is a possibility that every **R** allele from one parent fuses with an **R** allele from the other parent, upsetting the expected ratio of resulting phenotypes.

The greater the number of fertilisations involved, then the more likely it is that the resulting ratios will be close to the predicted. Just like tossing a coin, the greater the number of attempts then the more likely it will be that the results will be 50% heads and 50% tails.

ONLINE TEST

Check how well you've learned about inheritance online at www. brightredbooks.net/ N5Biology

THINGS TO DO AND THINK ABOUT

Use a Punnett square to predict the phenotype ratios of the offspring from the following crosses:

1   a heterozygous purple-flowered plant with a white-flowered plant

2   a heterozygous purple-flowered plant with a homozygous purple-flowered plant.

TRANSPORT SYSTEMS IN PLANTS 1

LEAF STRUCTURES

The main organs of plants are the roots, stems and leaves. All are involved in the transport of materials through the plant.

The structure of a typical leaf is shown in the diagram.

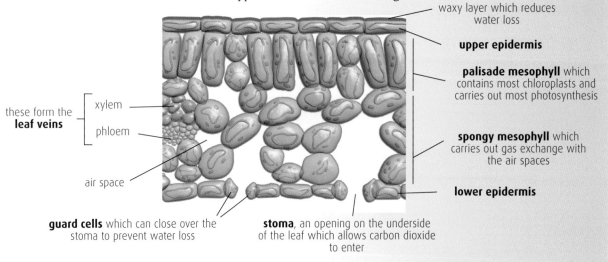

waxy layer which reduces water loss

upper epidermis

palisade mesophyll which contains most chloroplasts and carries out most photosynthesis

spongy mesophyll which carries out gas exchange with the air spaces

lower epidermis

these form the **leaf veins**

xylem

phloem

air space

guard cells which can close over the stoma to prevent water loss

stoma, an opening on the underside of the leaf which allows carbon dioxide to enter

THE TRANSPORT OF WATER AND MINERALS

There is a continuous movement of water from the roots of a plant to the leaves. The movement is always in that one direction and it is responsible for supplying plant cells with water and minerals. This movement of water involves a number of structures and processes.

The entry of water into plants

Plants obtain water and minerals from the soil through structures called root hairs. Root hairs can be seen on germinating seedlings such as radish. Each root hair is part of a specialised cell.

The root hairs come into close contact with the layer of water which surrounds soil particles. They provide a large surface area for the efficient uptake of water and minerals. Water enters root hair cells by osmosis. Minerals enter these cells by active transport.

Key
→ *water movement by osmosis*
→ *water movement into xylem vessels*

Xylem vessel

Root cell

Root hair

Water entering from soil

The movement of water from cell to cell in plants

Once water has entered a plant, it moves through the cells of the root by osmosis. The root hair cells have a higher water concentration than cells further inside the root because of the water they have absorbed from the soil, so there is a continual movement of water by osmosis across the root from the outermost cells to the inner cells adjacent to the xylem vessels in the centre of the root. Water movement in the xylem is always in the same direction, from the roots to the leaves. Minerals are carried in solution in the xylem. It is from the xylem that cells throughout the plant receive both the water and minerals they require.

contd

Water movement in xylem vessels

Xylem vessels are formed from elongated cells which have died, losing their contents and their end cell walls. The remaining cell walls become strengthened with deposits of **lignin** in characteristic patterns. The lignin is deposited on the inside of the cell walls in rings or spirals. Gradually, the deposits of lignin become more widespread until most of the inner walls of the xylem vessel are covered. The lignin enables the xylem vessels to withstand the pressure changes as water moves through the plant. The dead cells of the xylem form continuous hollow tubes which run from the roots of a plant, up the stem and into the leaves.

The upwards movement of water in plants can be demonstrated by placing a stalk of celery in coloured water. The colour eventually appears in the leaves of the celery, showing that it has been transported upwards from the base of the stalk. If the celery is cut, the location of the xylem vessels can be clearly seen.

Patterns of developing lignin deposits in xylem

Water Movements in Plant leaves

There is continual movement of water by osmosis in the leaves of a plant. Water evaporates from mesophyll cells of the leaf into the leaf air spaces. The cells which lose water have a lower water concentration than other cells. Osmosis takes place from cells which obtain water from xylem vessels to the cells which have lost water.

The water which evaporates from the mesophyll cells diffuses to the outside air through pores called **stomata** (singular – stoma) which are found in the lower epidermis of plant leaves. This loss of water from plant leaves is called **transpiration** and it is the main cause of the movement of water up the xylem vessels in the plant.

The overall movement of water through a plant is known as the transpiration stream. It is represented in the following diagram by the blue arrows.

Water movement in plant leaves

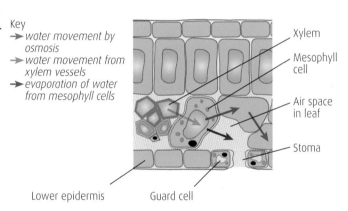

Key
→ water movement by osmosis
→ water movement from xylem vessels
→ evaporation of water from mesophyll cells

Xylem
Mesophyll cell
Air space in leaf
Stoma

Lower epidermis   Guard cell

Water movement in leaf

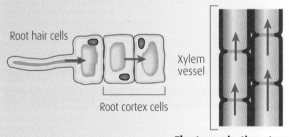

Root hair cells

Xylem vessel

Root cortex cells

The transpiration stream

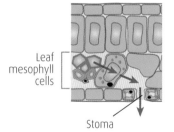

Leaf mesophyll cells

Stoma

 THINGS TO DO AND THINK ABOUT

Apart from the transport of water and dissolved minerals, suggest one other benefit a plant gains from having vessels strengthened with lignin running up its length from roots to leaves.

TRANSPORT SYSTEMS IN PLANTS 2

FACTORS AFFECTING TRANSPIRATION

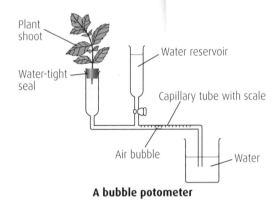

A bubble potometer

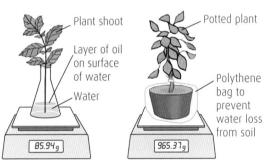

Weight potometers

As stated earlier, transpiration involves evaporation of water from the mesophyll cells in the leaves of plants. This means transpiration is affected by the same abiotic factors that influence evaporation. The rate of evaporation from plants and the effect of these factors can be measured using a piece of apparatus called a potometer. Diagrams of two types of potometer are shown below.

Bubble potometer

Bubble potometers measure the rate at which water is absorbed by the plant shoot. When water is lost by transpiration from the leaves, more water moves from the leaf xylem into the leaf mesophyll cells. This causes upwards movement of water in the xylem vessels, causing water to be drawn into the bottom ends of the xylem vessels.

The rate of this uptake of water can be measured by the rate of movement of the air bubble along the capillary tube. The water reservoir on the potometer allows the air bubble to be reset so that repeated measurements can be taken.

Weight potometer

Weight potometers measure the rate at which water is lost from the leaves of the plant or plant shoot. This is done by recording the weight of the plant or the shoot over a period of time. It is important that only water lost from the leaves is measured. To make sure of this, if a plant shoot is being used, a layer of oil is put on the surface of the water in the container to prevent evaporation.

If a whole plant is used, a polythene bag is secured around the stem of the plant and its pot for the same purpose.

Once the rate of transpiration from a plant has been measured several times and an average value calculated, the effects of various abiotic factors can be investigated by changing the conditions and taking more measurements. Details are shown in the table.

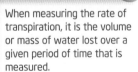
Factor tested	How the factor is altered	Effect on the rate of transpiration
Wind speed	Place a fan close to the apparatus	Increased wind speed increases transpiration
Humidity	Enclose the whole apparatus in a transparent container	Increased humidity decreases transpiration
Temperature	Move the apparatus to a warmer or cooler situation	Increased temperature increases transpiration
Surface area	Remove some leaves from the plant	Decreased total surface area decreases transpiration
Light	Place additional lamps close to the apparatus (use a heat shield to avoid increasing the temperature)	Increased light increases transpiration

The effects of changes in temperature, humidity and air movement on the rate of transpiration are to be expected, since these factors affect evaporation in the same way. However, changes in light intensity do not affect evaporation but do affect transpiration. This is because of the response of the **guard cells** surrounding the stomata to changes in light intensity. Stomata are open when it is light and closed when it is dark.

STOMATA

Stomata are the tiny pores found in the epidermis of plant leaves, mainly on the lower leaf surface. They are important in allowing carbon dioxide to enter the air spaces, so that it can be absorbed and used in photosynthesis. The number and shape of stomata vary from one plant species to another. Typically there are between 100 and 1000 stomata per square millimetre of leaf surface area.

Each **stoma** is surrounded by two specialised guard cells. The shape and structure of the guard cells cause them to bend when they are turgid. This opens a gap between them. The gap is the open stoma. When the guard cells are flaccid, they become closer together and so the stoma is closed.

Stomata from two different plants

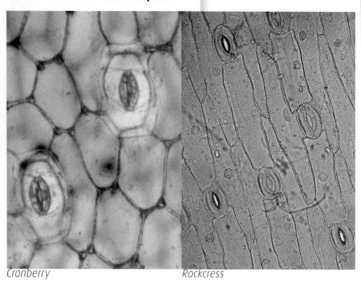

Cranberry   Rockcress

Water vapour passes out of the stomata during transpiration. The ability of the guard cells to open and close the stomata is important because it allows the stomata to close at night when photosynthesis does not take place. This reduces water loss from the plant at a time when it has no requirement for carbon dioxide uptake.

The appearance and relative number of stomata differ from species to species.

THE TRANSPORT OF SUGAR

Sugar is produced in the leaves of plants by photosynthesis. Plants use sugar for a number of different functions.

sugar
- used in respiration to release energy for use by cells
- converted to cellulose for the formation of cell walls
- converted to starch for storage
- used in the manufacture of other substances such as proteins and fats

This means that sugar is transported in a plant according to its requirements at any given time. Transport of sugar can be in any direction.

The transport of sugar takes place in phloem tissue. Like xylem, phloem tissue is formed from elongated cells which become arranged into vessels. However, the cells which form phloem vessels are still living, although they are highly adapted to enable them to transport material efficiently.

Sieve plate
(perforated end cell walls)

Cytoplasm strands
(running between sieve tube cells)

Companion cell
(its nucleus controls the sieve tube cell)

Phloem vessels

VIDEO LINK

For more on stomata, watch the video at www.brightredbooks.net/N5Biology

ONLINE TEST

Test yourself on transpiration and transport systems at www.brightredbooks.net/N5Biology

DON'T FORGET

Xylem carries water and minerals in only one direction. Phloem carries sugars in any direction.

THINGS TO DO AND THINK ABOUT

1 Why are the stomata found mostly on the lower surface of leaves?

2 What type of plant leaves have stomata only on the upper surface?

3 In a potato plant, suggest two reasons why sugar sometimes moves downwards from the leaves to underground parts of the plant and one reason why it moves upwards from underground parts.

TRANSPORT SYSTEMS IN ANIMALS 1

BLOOD CELLS

Blood consists of two parts:

- liquid plasma which contains dissolved materials
- blood cells.

Both the plasma and the red blood cells are important for the transport of materials around the body.

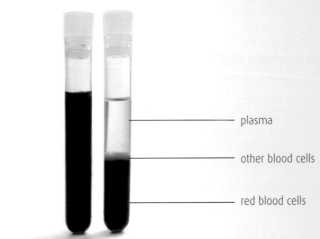

— plasma

— other blood cells

— red blood cells

Unseparated and separated blood

BLOOD CELLS

Red blood cells are the most numerous type of cell in our blood. Approximately one quarter of all our body cells are red blood cells. Our blood contains about 6 million red blood cells per cubic millimetre (mm³). A red blood cell lasts for about 120 days before it is broken down and replaced. Our bodies produce about 2·4 million new red blood cells every second.

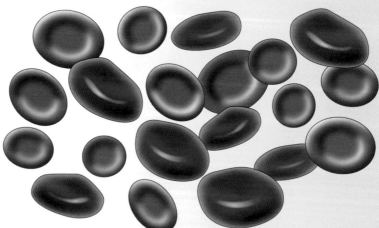

Red blood cells

The role of the blood circulatory system in the transport of hormones that are involved in the control and regulation of body processes has been mentioned earlier. In addition to the transport of hormones, blood is used to transport nutrients, oxygen, carbon dioxide and urea. Some details of these transport functions of blood are shown in the table.

contd

Red blood cells are very small but, even so, they can only pass through the smallest blood vessels, the capillaries, in single file. They are described as being biconcave in shape. This means they are flattened discs that are thinner in the middle than at the edge.

The function of red blood cells is to carry oxygen from the lungs to the body cells. They are able to do this because they contain a chemical called haemoglobin, which reacts with oxygen to form a temporary complex molecule called oxyhaemoglobin. At the lungs, oxygen from the air reacts with the haemoglobin and is carried by the red blood cells. At the body tissues, the oxyhaemoglobin releases the oxygen which is then available to the cells of the body.

This can be represented as a chemical equation.

in the blood capillaries of the lungs

oxygen + haemoglobin ⟶ oxyhaemoglobin

in the blood capillaries of the body tissues

Red blood cells do not have a nucleus. This is the reason why they have a short life span and need to be replaced.

White blood cells are larger and less numerous than red cells. They are part of the immune system and their function is to protect the body from infection from pathogens such as bacteria, viruses and fungi.

There are two main types of white cells:

- Phagocytes – these cells destroy pathogens by engulfing them and then destroying them using digestive enzymes. This process is called phagocytosis. Phagocytes can pass out of the blood vessels and actively move towards pathogens to destroy them.

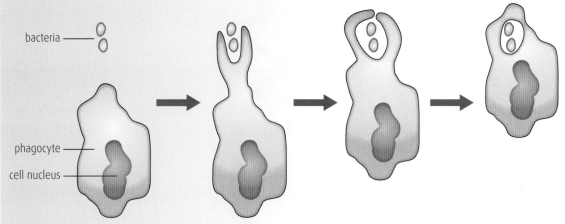

bacteria

phagocyte

cell nucleus

Phagocytosis

- Lymphocytes – these cells destroy pathogens by producing chemicals called antibodies. Pathogens contain chemicals called antigens which are recognised as being foreign to the body. When an antigen is recognised, a lymphocyte with an antibody specific to a particular antigen, will bind to the pathogen and destroy it. Once the body has made an antibody for a particular antigen, it is able to make the same antibody again very quickly. This means that if the same pathogen infects the body again, it will be rapidly destroyed. This means that you are immune to it.

lymphocyte

ONLINE TEST

Test yourself on transport functions of blood at www.brightredbooks.net/N5Biology

VIDEO LINK

Watch the clip on how oxygen is transported in blood at www.brightredbooks.net/N5Biology

DON'T FORGET

Red blood cells differ from other body cells because they do not have a nucleus.

DON'T FORGET

Red blood cells transport oxygen round the body. White blood cells protect the body from infection.

THINGS TO DO AND THINK ABOUT

Suggest how the lack of a nucleus and the distinctive shape of a red blood cell both contribute to its efficiency in transporting oxygen from the lungs to the cells of the body.

TRANSPORT SYSTEMS IN ANIMALS 2

THE STRUCTURE OF THE CIRCULATORY SYSTEM

Blood is pumped round the body by the muscular heart. The blood passes through the heart twice in one complete circuit. This is important because it means that the blood pressure remains high as the blood carries its contents to the body cells. One loop of the circuit takes the blood from the heart to the lungs and back to the heart. In the lungs the blood gains oxygen and loses waste carbon dioxide. The second loop of the system takes the freshly oxygenated blood from the heart round the rest of the body and back to the heart. As the blood travels round this loop, the body cells gain oxygen and nutrients, and get rid of carbon dioxide.

Heart structure

The following diagram shows the structure of a dissected heart and its major blood vessels.

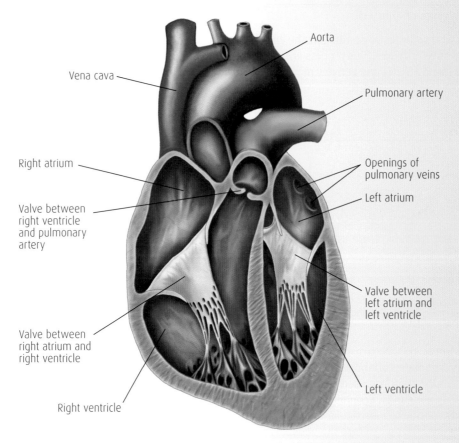

Aorta

Vena cava

Pulmonary artery

Right atrium

Openings of pulmonary veins

Left atrium

Valve between right ventricle and pulmonary artery

Valve between left atrium and left ventricle

Valve between right atrium and right ventricle

Left ventricle

Right ventricle

Dissected heart

Notice the much thicker muscular wall of the left **ventricle** compared to the right ventricle. This is because the left ventricle has to pump blood round the whole of the body, whereas the right ventricle pumps blood only to the lungs.

The heart contains four **valves** which prevent the blood from flowing backwards during the heart contractions. There is a valve between the **atrium** and the ventricle on each side of the heart and a valve between the ventricle and the main **artery** which leaves the ventricle. In the diagram, the valve between the atrium and the ventricle on each side of the heart can be seen, as can the valve between the right ventricle and the **pulmonary artery**.

contd

The pathway of blood round the body

The following diagram shows the route taken by the blood around the body.

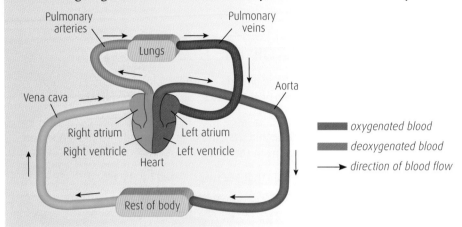

DON'T FORGET

1 Arteries always carry blood away from the heart. Veins always carry blood towards the heart.
2 The atria (singular atrium) are the receiving chambers of the heart. The ventricles are the pumping chambers.
3 The left ventricle has a thicker muscular wall than the right ventricle because it needs to pump the blood further than the right ventricle.
4 Diagrams showing body structures are always presented as though the organism is facing you. This means that the right-hand side of the diagram represents the left-hand side of the organism.

BLOOD VESSELS

Blood is carried round the body in three different types of blood vessel. The different blood vessels have different structures to match their different functions.

Arteries

Arteries carry blood away from the heart. This blood is under high pressure and so arteries have a relatively narrow channel and thick muscular walls to help maintain the pressure of the blood as it travels round the body.

The pulmonary arteries carry deoxygenated blood from the right ventricle to the lungs where it gains oxygen.

The largest artery is the aorta which carries oxygenated blood from the left ventricle. Smaller arteries branch from the aorta to carry the blood to particular areas and organs.

The first branch from the aorta is the **coronary artery**. This is located within the wall of the heart and supplies the heart muscle with nutrients and oxygen.

Veins

Veins carry blood towards the heart. The blood is returning from tissues and organs of the body. The **pulmonary veins** from the lungs carry oxygenated blood to the left atrium of the heart, so that it can be pumped to the rest of the body. Veins from these other parts of the body feed into the largest vein, called the **vena cava**, which takes the deoxygenated blood to the right atrium.

The blood in the veins is at low pressure and so veins have a thinner muscle layer in their walls and a wider channel than the arteries. This means that the veins do not create much resistance to the flow of blood. Veins also have valves at intervals along their length so that the slow-moving blood does not begin to flow backwards.

Capillaries

Capillaries are the blood vessels which link the arteries and veins. They form a network of vessels in the tissues of the body. Capillaries are microscopic in size and have very thin walls, only one cell thick at their finest. A network of capillaries produces a large surface area for the efficient exchange of materials between the blood and the body cells.

VIDEO LINK

Watch the clip for a video on capillaries at www.brightredbooks.net/N5Biology

Cross-sections of blood vessels

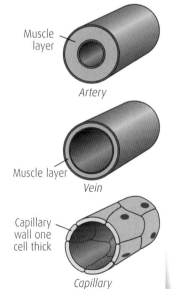

Muscle layer

Artery

Muscle layer

Vein

Capillary wall one cell thick

Capillary

ONLINE TEST

Test yourself on the structure of the circulatory system at www.brightredbooks.net/N5Biology

THINGS TO DO AND THINK ABOUT

Using the names of the major blood vessels and the chambers of the heart, describe the path of a red blood cell from the time it leaves the lungs until it returns back to the lungs.

ABSORPTION OF MATERIALS 1

DON'T FORGET

1 Red blood cells transport oxygen and a little carbon dioxide. All other materials are transported in solution by the plasma.
2 In addition to the transport of materials, the blood is also important in distributing heat around the body.

TRANSPORT FUNCTIONS OF THE BLOOD

Oxygen and nutrients from food must be absorbed into the bloodstream. They must then pass from the blood into the cells of the body to be used for respiration, protein synthesis and other processes.

Waste materials such as carbon dioxide and urea must be absorbed into the blood and later pass from the blood for removal from the body.

The table shows the variety of materials that are transported by the blood.

Material	Where it enters the blood	Where it leaves the blood	How it is transported	Use in the body
Oxygen	Lungs	Body cells	In the red blood cells as oxyhaemoglobin	Used in respiration to release energy from glucose
Glucose	Small intestine	Body cells	In solution in the plasma	Used as an energy source and for making other chemicals. Excess can be stored as glycogen or fat.
Amino acids	Small intestine	Body cells	In solution in the plasma	Used to make proteins. Excess is broken down in the liver to make urea.
Carbon dioxide	Body cells	Lungs	In solution in the plasma	Waste material produced during respiration.
Urea	Liver	Kidneys	In solution in the plasma	Waste material produced from excess amino acids.
Hormones	Various endocrine glands	Affect specific target cells	In solution in the plasma	Regulation of various body functions.

VIDEO LINK

Watch the clip at www. brightredbooks.net to learn more about capillaries.

CAPILLARY NETWORKS

The exchanges of materials between the blood and body cells is possible because of the network of capillary blood vessels which provides every cell of the body access to the blood system.

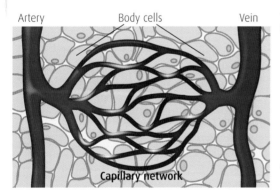

Artery   Body cells   Vein

Capillary network

At every point where these materials enter or leave the blood, the surfaces across which the materials pass show some common features:

- Large surface area of contact between the blood capillary and the cell from which, or to which, the material is passing.

- Thin walls to allow rapid diffusion of the material. The walls of the capillary vessels are only one cell thick.

- Extensive blood supply. Capillary vessels are so small and so numerous that every body cell has contact with them.

These features increase the efficiency of the exchange of materials.

GAS EXCHANGE IN THE LUNGS

Breathing

Air passages in the nose and mouth are connected to the trachea. This divides into two bronchi, one going into each lung. Each **bronchus** divides repeatedly into smaller and smaller air tubes called **bronchioles**. The trachea is strengthened by rings of cartilage which prevents it from kinking when moving the neck. Cartilage rings are also present in the bronchi and the larger bronchioles to prevent these air passages from collapsing during inhalation.

contd

Finally the bronchioles end in microscopic bubble-like air sacs called alveoli.

The breathing system

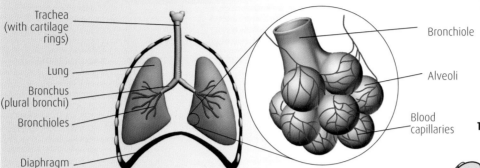

Trachea (with cartilage rings)

Lung

Bronchus (plural bronchi)

Bronchioles

Diaphragm

Bronchiole

Alveoli

Blood capillaries

Before gas exchange can take place, air must enter the lungs from the atmosphere. This is achieved by the actions of the ribs and the diaphragm. The ribs are connected to the bones of the spine at the back and to the breast bone at the front. Together these form a protective cage-like structure which surrounds the lungs. The diaphragm is a muscular sheet underneath the lungs. Movement of the ribs and the diaphragm increase and decrease the volume of the lungs. This forces air in and out.

GAS EXCHANGE

The alveoli of the lungs are microscopic. They are bubble-like structures with walls that are only one cell thick. Together they provide a very large surface area for gas exchange. Capillary blood vessels form a network over the outside of the alveoli. Gas exchange takes place between the blood in the capillaries and the air in the alveoli.

There is a higher concentration of oxygen in the alveolar air than in the blood, so oxygen diffuses from the air into the blood. It has only to pass through the wall of the **alveolus** and the wall of the blood capillary, both just one cell thick. At the same time, carbon dioxide diffuses in the opposite direction because the blood has a higher concentration of carbon dioxide than air.

The difference in the concentrations of each of the two gases in the air compared to their concentrations in the blood is always maintained. This is because the air in the alveoli is always being replaced with air from the outside and fresh blood is always flowing through the capillaries.

The large number of alveoli and the extensive network of blood capillaries which surround them create a large surface area for the diffusion of gases between the blood and the air in the alveoli.

The mechanism of breathing

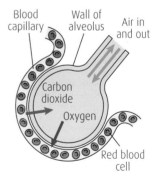

Breathing in

Rib cage moves upwards and outwards

Diaphragm moves downwards

Breathing out

Rib cage moves downwards and inwards

Diaphragm moves upwards

Gas exchange in an alveolus

Blood capillary

Wall of alveolus

Air in and out

Carbon dioxide

Oxygen

Red blood cell

 THINGS TO DO AND THINK ABOUT

1. Efficient absorption of material into cells requires (i) a large surface area, (ii) thin walls and (iii) a good blood supply.

   Explain how a capillary network satisfies these requirements.

2. What is the equation for the reaction between oxygen and haemoglobin?

3. Describe the features of red blood cells which make them efficient in transporting oxygen.

ABSORPTION OF MATERIALS 2

THE DIGESTION SYSTEM

The digestive system is responsible for the breakdown of large insoluble molecules in our food to smaller soluble molecules, which can then be absorbed into the blood. The overall reactions of digestion can be summarised as:

1 complex carbohydrates → glucose

2 proteins → amino acids

3 fats → fatty acids and glycerol

The reactions of digestion take place in the mouth, the stomach and the first half of the small intestine. Enzymes break down food particles into their products. The products are then absorbed into the blood. This takes place from the second half of the small intestine.

The various parts of the digestive system and some associated organs are shown in the diagram.

DON'T FORGET

The purpose of digestion is to breakdown large insoluble food molecules to smaller soluble molecules so that they can be absorbed into the blood.

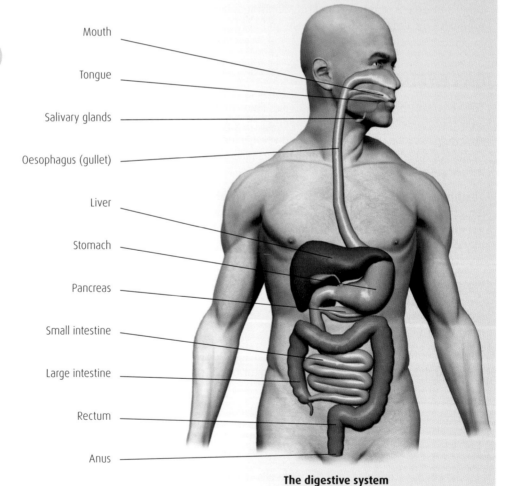

Mouth
Tongue
Salivary glands
Oesophagus (gullet)
Liver
Stomach
Pancreas
Small intestine
Large intestine
Rectum
Anus

The digestive system

ABSORPTION FROM THE SMALL INTESTINE

Digestion of food is complete by the time it reaches the second half of the small intestine. This part of the digestive system is adapted to absorb the digestion products into the blood. The inner lining of the small intestine is covered in small projections called **villi** (singular **villus**) that are about 1 mm in length. This gives the inner lining of the small intestine a greatly increased surface area, enabling the absorption of the digestion products to take place efficiently and quickly.

Structure of a villus

Villi have a number of features that help to absorb the products of digestion. These can be seen in the diagram.

Internal structure of a villus

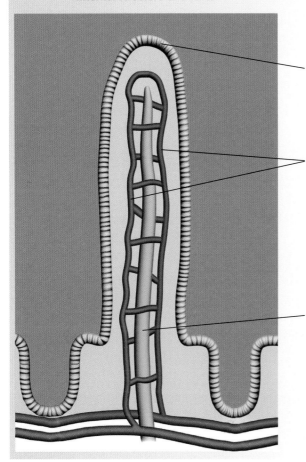

Thin wall (one cell thick) – enables digestion products to diffuse quickly from the small intestine to the transport vessels

Blood capillaries – absorb glucose and amino acids

Lacteal (lymph capillary) – absorbs the products of fat digestion (fatty acids and glycerol), which are less soluble than other digestion products

DON'T FORGET

A villus is a very small structure. There are a very large number of them which produce a very large total surface area.

VIDEO LINK

For a tutorial about the digestive system, watch the video at www.brightredbooks.net/N5Biology

ONLINE TEST

For a test on digestion, visit www.brightredbooks.net/N5Biology

THINGS TO DO AND THINK ABOUT

1   From which part of the digestive system does the absorption of nutrients take place?

2   Explain how the villi satisfy the requirements for the efficient absorption of materials described on page 70.

LIFE ON EARTH

ECOSYSTEMS 1

ECOLOGICAL TERMS

Ecology is the study of organisms in their environment. There are some ecological terms which are used frequently and which should be learned. The definitions of the more common ones are given in the table.

Term	Definition
Species	A group of organisms which are able to interbreed to produce fertile offspring
Biodiversity	The range of species present in an ecosystem. Biodiversity can be considered at other levels, for example the whole planet
Population	All the members of a single species in a given area
Community	All the organisms, plants and animals, which are present in a habitat.
Producer	Green plants which make their own food by photosynthesis
Consumer	An organism which gets its energy by feeding on other organisms, their remains or their waste
Herbivore	An animal which gets its energy by eating plants
Carnivore	An animal which gets its energy by eating animals
Omnivore	An animal which gets its energy by eating both plants and animals
Predator	An animal which kills other animals for food
Prey	An animal which is killed and eaten by another animal
Food chain	A sequence of organisms showing the feeding relationships between them.
Food web	The feeding relationships of all the organisms present in a habitat. A food web normally contains numerous interconnected food chains

DON'T FORGET

Decomposers such as bacteria and fungi play an important part in any ecosystem. They are responsible for the breakdown of dead organic matter which allows mineral nutrients to become available for new plant growth.

VIDEO LINK

Check out the clip 'What is an ecosystem?' from www.brightredbooks.net/N5Biology

Ecosystems

An ecosystem consists of all the organisms (the community) living in a particular habitat and the non-living components (abiotic factors) with which the organisms interact.

Niche

A **niche** is the role that an organism plays within its community. A description of a niche would include the range of food sources the organism uses, the range of predators which feed on it, the range of other species it competes with and the habitat it occupies.

For example, the niche of the red fox is that of a predator, active at night and feeding on small mammals, amphibians, insects and fruit. The fox provides blood for blackflies and midges, and is host to numerous diseases. The scraps, or carrion, left behind after a fox's meal provide food for many small scavengers and decomposers. All these are found in the habitat of woodland and meadow.

DON'T FORGET

The niches of different species in a community may overlap, but they are not identical.

FOOD CHAINS

Living organisms obtain the energy they require from their food. Some organisms, namely green plants, make their food by photosynthesis using energy from the Sun. All other organisms obtain their food from the bodies, or the remains, of other organisms.

The transfer of energy between organisms is often represented as a **food chain**.

As shown in the diagram, each stage of a food chain has a name.

VIDEO LINK

Watch the clip 'Energy transfers and food chains' for more on this at www.brightredbooks.net/N5Biology

A simple food chain

secondary consumers

producer *primary consumer*

grass plant-eating insect spider blackbird fox

- A **producer** is a green plant which makes food by photosynthesis.

- A primary **consumer** is a **herbivore** – an animal which eats plants.

- Secondary consumers are **carnivores** – animals which eat other animals.

The arrows of a food chain represent the direction of energy transfer from organism to organism.

The transfer of energy from one organism to the next is not complete because each organism uses up some of the energy, so there is less energy remaining to be passed to the next stage. Some of the energy obtained by an organism remains in the organism, for example energy is present in new tissues when the organism grows. This will be available to the next stage in the food chain if the organism is eaten. However, energy which is converted to heat or which is used for movement will be lost from the food chain and will be unavailable to the next stage. Some energy will remain in uneaten material such as bones, teeth, fur and feathers. There will also be energy remaining in the undigested material present in faeces. This energy will not be passed to the next stage, although it may be used by decomposers such as bacteria and fungi that are present in the ecosystem.

ONLINE TEST

Check out the clip 'What is an ecosystem?' from www.brightredbooks.net/N5Biology

THINGS TO DO AND THINK ABOUT

Describe the niche of an organism based on a food web. In your description, make sure you account for every arrow pointing to and pointing away from the organism and add any information you have about the habitat.

DON'T FORGET

The energy for all organisms comes from the Sun. A food chain must start with a producer because they are the only organisms able to use light energy from the Sun to make food. It is through the producers that energy is made available to all the other organisms present. All the other organisms are consumers.

ECOSYSTEMS 2

FOOD WEBS

In reality, the transfer of energy in an ecosystem is not as simple as it appears in a food chain. This is because most organisms have more than one food source and any one type of organism may be eaten by a range of predators. Therefore, most food chains are interlinked into more complicated systems called food webs.

For example, part of a food web involving the organisms from the food chain is shown below.

A food web

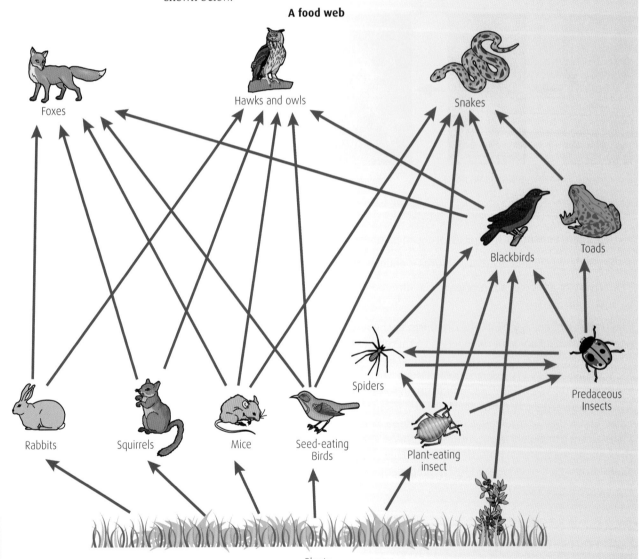

Foxes · Hawks and owls · Snakes · Blackbirds · Toads · Spiders · Predaceous Insects · Rabbits · Squirrels · Mice · Seed-eating Birds · Plant-eating insect · Plants

ONLINE TEST

Test yourself on Ecosystems at www.brightredbooks.net/N5Biology

A description of the position that an organism occupies in a food web is very close to a description of its niche.

It is possible to find many individual food chains in a food web such as this.

EFFECTS OF THE LOSS OF ORGANISMS FROM A FOOD WEB

Food webs can be affected by changes in the population of any of the organisms which are part of it. For example, in the food web on page 76, if the aphid population was to fall then there would be a decrease in the number of spiders and predaceous insects. This would cause a decrease in the blackbird and toad populations which, in turn, would affect the populations of foxes, hawks and owls, and snakes. There would be further consequences for the populations of rabbits, squirrels, mice and seed-eating birds.

The greater the complexity or biodiversity of a food web, the more stable it is likely to be. This means that it will be able to adapt to changes caused by the loss of some organisms.

Consider the Canadian food web on the right:

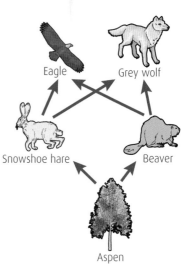

If the beaver population fell because of a fatal disease, then a major food source of eagles and wolves would be lost. With hares then being the only food source for the eagles and wolves, the population of hares would soon become too small to support the eagles and wolves. Therefore the whole food web would have collapsed.

Compare this situation to the more complex food web below.

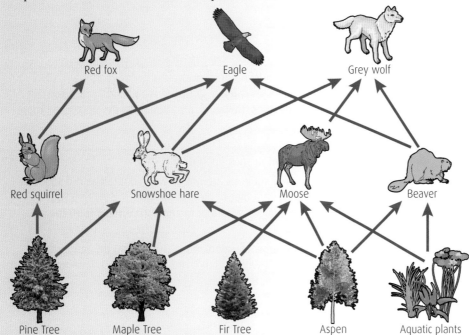

In this case, the loss of the beavers would not have such a catastrophic effect because the eagles and wolves still have several other food sources available. The extra demand placed on these other food sources would cause some fluctuations in numbers but in time a new balance would be achieved amongst the remaining populations.

VIDEO LINK

Watch the video at www. brightredbooks.net to learn more about ecosystems.

DON'T FORGET

The greater the biodiversity (number of different species present) in a food web, the greater is its ability to adapt to changes caused by the loss of one species.

THINGS TO DO AND THINK ABOUT

From the food web on page 76:

(a) name all the herbivores

(b) name all the omnivores

(c) name all the carnivores

(d) name all the animals which are both predator and prey.

ECOSYSTEMS 3

COMPETITION

Competition results in ecosystems when organisms require the same resources for their survival and these resources are in short supply. These resources include food, water, light, territory and breeding partners. Competition does not mean that individuals fight over the resource, although this sometimes happens. It does mean that some organisms receive less of the resource than others. These individuals have a reduced chance of surviving or of reproducing successfully.

INTERSPECIFIC COMPETITION

Interspecific competition occurs when individuals of different species in an ecosystem require similar resources. Whenever this happens, one species will prove to be a stronger competitor than the other. This results in the population of the weaker species declining in the ecosystem and perhaps being eliminated from it.

The distribution of red squirrels and grey squirrels 1998

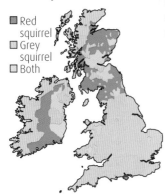

☐ Red squirrel
☐ Grey squirrel
☐ Both

Red squirrel   **Grey squirrel**

EXAMPLE 1

Red squirrels and grey squirrels

Red squirrels are the native squirrel species in Britain. Grey squirrels were introduced into Britain from North America over a period of years from 1876 as a novelty species. Both species can live in mixed woodlands which contain broad-leaved and coniferous trees. However, grey squirrels are bigger and more robust than red squirrels.

They are stronger competitors and populations of red squirrels are eliminated when both species are present in such habitats.

In coniferous woodlands, grey squirrels are not as well adapted as red squirrels because they cannot survive on a diet of pine seeds. This means that red squirrels can survive in some areas where pine and spruce trees form the predominant vegetation.

The overall reduction in the areas of woodland in Britain and the tendency for replacing coniferous woodlands with broadleaved tree species has contributed to the decline of red squirrels in Britain.

EXAMPLE 2

Brown trout and rainbow trout

Brown trout are a native freshwater fish. Rainbow trout are a different species, introduced from the USA for angling and as a food.

Brown trout   **Rainbow trout**

Both species lay eggs in hollows, or redds, scraped in the gravel of a river bed. Research has shown that when both species use the same areas of a river bed for redding, the eggs of the brown trout suffer a greater mortality rate than those of the rainbow trout. This means that the population of brown trout declines in rivers where both species are present. In the UK, rainbow trout are restricted to fish farms or fisheries although there are reports of some wild populations.

INTRASPECIFIC COMPETITION

Intraspecific competition occurs between members of the same species. It can be very intense because the requirements of all the individuals involved are identical. Competitive success is based on variations between the individuals. It ensures that the fittest and best adapted have the greatest chance of surviving and reproducing.

EXAMPLE 1

Territorial behaviour in robins

Robins are very territorial birds. Each male robin defends his territory against other robins to ensure that he controls an area with sufficient food for himself and his offspring. The red breast is a warning to other robins to stay out of the defended territory.

Posturing and singing are usually enough to deter other robins and to prevent actual fighting. However, if these warnings are ignored, then a robin will attack an intruder in order to prevent a competitor from obtaining the resources available in the territory.

EXAMPLE 2

Grasshoppers competing for food supply

Grasshoppers eat the leaves of plants. They do not challenge each other for the food but simply compete by eating. By doing so, they reduce the food available to their competitors.

EXAMPLE 3

Trees growing close together compete for water, light, nutrients and space

Competition for light encourages trees to grow tall. A single tree may not grow as tall as trees of the same species which are growing close together, but it will have more resources and be wider and heavier than a tree from a group.

 DON'T FORGET

Competition can occur between different species or between individuals of the same species – whenever two organisms require the same resources there will be competition for that resource.

 THINGS TO DO AND THINK ABOUT

The diagram shows the set up for an investigation into competition between radish seedlings.

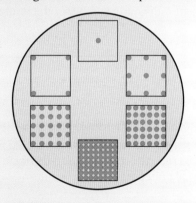

Petri dish with wet filter paper on the bottom

1 cm² areas containing different numbers of radish seeds evenly spaced out

1   By using one dish for all the samples, what factors are being kept the same?

2   How can you measure the effect of increasing competition on the growth of the seedlings?

3   Is the average height of the seedlings in each sample a suitable measure?

4   What would be a better measure of seedling growth?

5   How could the results be made more reliable?

DISTRIBUTION OF ORGANISMS 1

MEASURING ABIOTIC FACTORS

Abiotic factors are the physical factors which characterise a habitat. The measurement of abiotic factors is important in understanding the distribution of organisms in the habitat.

Just as with the sampling of organisms, the measurement of abiotic factors must be representative and reliable. This means that care must be taken to avoid errors with the measurements and that the measurements must be repeated, so that average values can be calculated. This reduces the effect of any extreme or atypical measurements.

TEMPERATURE

Measurements are made in degrees Celsius (°C) using a thermometer or a temperature probe attached to a data recorder.

Thermometer

Soil thermometer

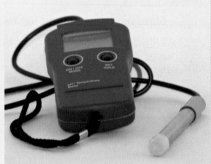

Data recorder and temperature probe

Whichever method you are using to record temperature, it is important to take the following precautions:

- Make sure the detecting part of the instrument is surrounded by the material being measured, whether it is water, air or soil.

- Avoid touching the detecting part of the instrument with your hand so that the results are not affected by your body heat.

- Allow the instrument to stabilise before you read the temperature.

LIGHT

Several different properties of light may be measured. Normally, it is the amount of light per unit of area, or light intensity, which is of interest. This is measured in lux (lx) using a light meter, which may be connected to a data recorder or can be a stand-alone instrument. Often light-measuring instruments are calibrated with arbitrary units of measurement.

When measuring light intensity, it is important to observe the following precautions:

- Make sure the detector is not dirty.

- Angle the detector so it is facing towards the source of light.

- Avoid shading the detector with your body or any other object.

DON'T FORGET ✚

To get accurate readings every time, ensure you do not alter the results by touching the thermometer bulb or blocking the light meter!

pH

pH is a measure of acidity or alkalinity. The pH scale goes from 0 to 14, with acids at the low end of the scale and alkalis at the high end. pH 7 is described as neutral – that is neither acidic nor alkaline. Strong acids and alkalis have values close to the extreme ends of the scale. Weak acids and alkalis have values close to the middle of the scale.

The pH of soil water is an important factor in determining which plants are able to survive in a habitat. The pH of waterways is discussed on page 89.

pH can be measured using indicator solutions which change colour at different pH values. For more accurate measurements a pH meter is used. This may be in the form of a probe attached to a data recorder or a stand-alone unit. Care must be taken to:

- ensure good contact between the probe and the material to be measured

- clean the probe between each measurement to make sure traces from a previous measurement do not affect the reading.

SOIL MOISTURE

This is another important factor which influences the distribution of plants. It is measured using a moisture meter. The format and precautions in the use of moisture meters are the same as for pH meters.

Some soil monitors combine several functions into a single unit.

A soil monitor combining light, soil pH and soil moisture meters.

 VIDEO LINK

Check out the video about water sampling to see how biologists use these methods of measurement at www.brightredbooks.net/N5Biology

 ONLINE TEST

Test how well you know your sampling and measurement techniques for abiotic factors at www.brightredbooks.net/N5Biology

 THINGS TO DO AND THINK ABOUT

The diagram shows the positions of 10 quadrats placed 1 metre apart along a transect. The transect runs from the bottom of a tree into open ground. (For information on transects, see page 83.)

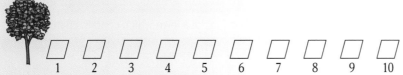

The table shows the number of daisy plants found in each quadrat, together with the values of some abiotic factors.

Quadrat	Number of daisies	Soil pH	Soil temperature (°C)	Soil moisture (units)	Light intensity (units)
1	0	6·5	14	4	8
2	1	6·4	15	6	8
3	1	6·8	14	8	9
4	3	6·7	15	12	10
5	4	6·5	16	9	13
6	6	6·4	15	7	15
7	9	6·7	15	11	16
8	10	6·6	15	11	16
9	9	6·5	14	10	18
10	10	6·7	15	8	18

1 Which of the abiotic factors is having the greatest influence on the distribution of the daisies?

2 Can you explain your answer?

DISTRIBUTION OF ORGANISMS 2

SAMPLING ORGANISMS

Quantitative sampling enables estimates to be made of the population sizes of the organisms in a habitat. Different techniques are available to suit different circumstances. In all cases, it is important that the samples are representative of the whole area and that enough samples are taken to make sure that the results are reliable.

This is usually achieved by selecting the sample sites randomly, so that the results are not influenced by any preconceived ideas of the investigator. The number of samples taken should be large enough to ensure that the effects of any extreme or atypical results are reduced when an average value is calculated.

Plants do not move and this makes sampling them easier.

SAMPLING PLANTS

Quadrats

A **quadrat** is a square frame, usually measuring 0·5 metre × 0·5 metre, giving it an area of 0·25 m².

If you wanted to compare the distribution of daisies on two different lawns, you could count every single daisy but this would be very time consuming. Instead, sample sites are chosen at random and a quadrat is placed at each site. The number of daisies in each quadrat is counted and an average is calculated. From this, the total number of daisies in the lawn can be estimated.

Quadrats can also be used to measure the proportion of the ground that is covered by plants which are too difficult to count individually, for example grasses and mosses.

To do this, the quadrat is divided into a grid of smaller squares using string. When the quadrat is positioned at the sample site, the number of small squares containing the plant is used as an estimate of the abundance of the plant in terms of the percentage ground cover.

The diagram shows a quadrat positioned on part of a lawn which contains patches of moss.

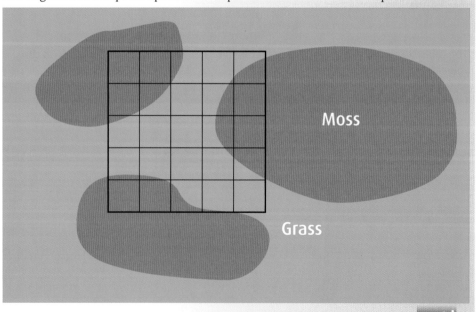

contd

Moss is present in 20 of the 25 squares of the quadrat. Some of these squares are completely filled with moss and some have hardly any. When using a quadrat to estimate ground cover, only the squares that are at least half-filled are counted.

In this case, 9 of the 25 small squares of the quadrat are at least half-filled with moss. This equals 36%. If this pattern was found to be the average of several sample sites, it could be concluded that 36% of the lawn was covered in moss.

Transects

Transects involve placing quadrats at regular intervals along a line. The results of samples can be used to study the effect of changes in an abiotic factor on the distribution of plants.

For example, a **transect** may be positioned down a slope to investigate the effect of increasing soil moisture from the top to the bottom of the slope. Another example could be to position the transect from a shaded area into open ground to investigate the effect of increasing light intensity. Read more about abiotic factors which affect ecosystems on page 89.

Read more about abiotic factors which affect ecosystems on page 89.

THINGS TO DO AND THINK ABOUT

The diagrams represent two lawns. The sites of quadrats are shown, together with the number of daisies found in each of them. Each quadrat is 0·25 m².

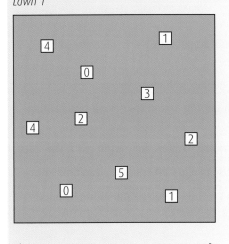

Lawn 1

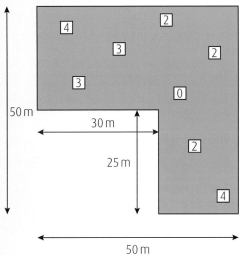

Lawn 2

Lawn 1

Total number of daisies in quadrats

= 4 + 1 + 0 + 2 + 3 + 4 + 2 + 0 + 5 + 1 = 22

Average number of daisies in quadrats

= 22 ÷ 10

= 2·2 per 0·25 m² = 8·8 per m²

Estimated total in Lawn 1 = 8·8 × 50 × 50 = 22 000

1 Estimate the total number of daisies in Lawn 2.

2 Which lawn contains the most daisies?

3 Which lawn has the greater density of daisies (number per unit area)?

DISTRIBUTION OF ORGANISMS 3

SAMPLING ANIMALS

Animals which do not move quickly, such as limpets on rocks at the seashore, may be sampled using a quadrat. For animals which move about, a method of trapping them is needed in order to take a sample.

Pitfall trap

The **pitfall trap** is the most common method used to trap small invertebrates. It consists of a container buried level with the soil surface and protected with a suitable raised cover. It is left for some time and then examined to see what animals have fallen in.

A pitfall trap

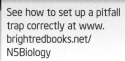

VIDEO LINK

See how to set up a pitfall trap correctly at www.brightredbooks.net/N5Biology

Some precautions must be taken when using pitfall traps if they are to give usable results:

- The rim must be level with the soil surface to enable animals to fall in.
- The trap must have a raised cover to stop rain getting in and to prevent birds eating any trapped animals.
- There must be a gap between the cover and the soil to allow animals to reach the trap.
- The container should have small holes in the base to allow any rain water to drain away.
- The trap must be checked frequently to collect results before some of the trapped animals eat others.
- A sufficient number of traps must be set to ensure representative and reliable results.

DON'T FORGET

When sampling animals, it is vital to check your traps regularly so that none of the animals in your sample are eaten by others, which may be predators.

An estimate of the population size for any of the trapped species can be made using a capture and recapture technique. This involves the following procedure:

1 Capture a sample of individuals of the species being studied and count them.

2 Mark each individual in some way, for example with a spot of waterproof ink on their backs.

3 Release them and re-set the traps to capture a second sample.

4 Count the total number of individuals in the second sample and also count how many of them were part of the first sample.

5 The estimated total population of the species is calculated as:

$$\frac{\text{number caught in first sample} \times \text{number caught in second sample}}{\text{number of marked individuals in second sample}}$$

ONLINE TEST

How well do you know your sampling and measurement techniques for biotic factors? Check at www.brightredbooks.net/N5Biology

contd

Tullgren funnel

This is used for sampling organisms in soil or in leaf litter. It consists of a funnel fitted with a mesh platform. The soil or leaf litter sample is placed on the platform and a lamp is positioned above the sample. Small animals move away from the heat and light of the lamp and fall through the mesh into a container placed under the funnel.

Only animals small enough to pass through the mesh will be collected and some may be killed before they escape the heat from the lamp.

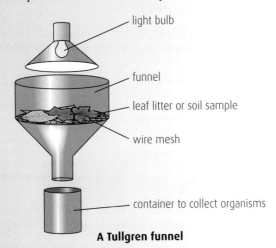

light bulb

funnel

leaf litter or soil sample

wire mesh

container to collect organisms

A Tullgren funnel

Tree beating

This is used to sample organisms that rest or feed in the branches or foliage of trees. As the branches are shaken, the organisms fall and are collected on a sheet.

Some animals may fly away or cling too tightly to be collected in the sample.

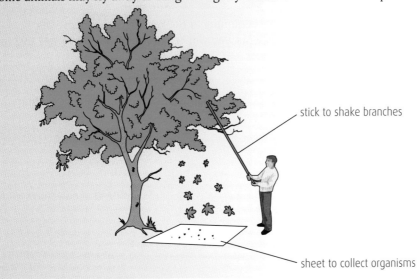

stick to shake branches

sheet to collect organisms

 THINGS TO DO AND THINK ABOUT

20 pitfall traps were set up overnight in an uncultivated area of a garden. The next morning the traps were found to contain a number of different organisms, including a total of 15 Devil's coach horse beetles. The beetles were marked and released and the traps were reset.

The next morning the traps contained a total of 12 Devil's coach horse beetles (see page 86), including 4 marked beetles.

Calculate the estimated number of Devil's coach horse beetles living in this area of the garden.

DISTRIBUTION OF ORGANISMS 4

IDENTIFYING ORGANISMS USING PAIRED-STATEMENT KEYS

You may want to identify animals or plants that have been found in a sample. This is normally carried out using a **paired-statement key**.

This key consists of a set of paired statements about features of organisms likely to be present in the habitat. The pairs of statements are organised in a sequence so that when the features of an unidentified organism are compared with the statements, they lead the scientist through the key until the organism is identified.

The photographs show some common invertebrates found in leaf litter. Note that the photographs *do not* show the organisms to the same scale.

The key can be used to identify them.

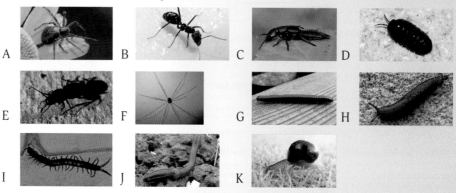

Look at photograph A and start at the beginning of the key. As you compare the features of the animal with the statements, you will be led from statement 1 to 2 to 3 to 4 to 7 to 8 and finally to the name Spider.

Key

Number	Paired statements	Organism name or further instruction
1	Shell present	Snail
	Shell not present	Go to 2
2	Body without segments	Slug
	Body with segments	Go to 3
3	Legs not present	Earthworm
	Legs present	Go to 4
4	Three pairs of legs	Go to 5
	More than three pairs of legs	Go to 7
5	Wings absent	Ant
	Wings present	Go to 6
6	Short wing covers and exposed abdomen	Devil's Coach Horse Beetle
	Long wing covers and covered abdomen	Ground Beetle
7	Four pairs of legs	Go to 8
	More than four pairs of legs	Go to 9
8	Body divided into two parts	Spider
	Body not divided into two parts	Harvestman
9	Body with fewer than 15 segments	Woodlouse
	Body with more than 15 segments	Go to 10
10	One pair of legs per segment	Centipede
	Two pairs of legs per segment	Millipede

CONSTRUCTING A PAIRED-STATEMENT KEY

To make a paired-statement key for others to use to identify organisms, information about easily observed features must be gathered.

Some features of six plants which belong to the buttercup family are shown in the table.

Plant name	Leaf shape	Runners (horizontal stems)	Stem
Greater spearwort	toothed	present	hairy
Meadow buttercup	lobed	absent	hairy
Lesser celandine	heart-shaped	absent	hairless
Creeping buttercup	lobed	present	hairy
Lesser spearwort	toothed	absent	hairless
Celery-leafed buttercup	lobed	absent	hairless

To construct the key, choose one of the features which could be used to divide the plants into two groups. Write a pair of opposite statements which does this.

For each group give an instruction to go to another pair of statements about a different feature which divides the group into two smaller groups.

Continue this until only one plant is in each group. At this point the plant can be named. Other people will then be able to take one of the plants, which they do not know, and follow the key until they come to its name.

1	Leaves toothed-shaped	....	Go to **2**
	Leaves not toothed-shaped	....	Go to **3**
2	Runners present	....	Greater spearwort
	Runners absent	....	Lesser spearwort
3	Hairy stem	....	Go to **4**
	Hairless stem	....	Go to **5**
4	Runners present	....	Creeping buttercup
	Runners absent	....	Meadow buttercup
5	Leaves heart-shaped	....	Lesser celandine
	Leaves lobed	....	Celery-leaved buttercup

THINGS TO DO AND THINK ABOUT

Use the key for leaf litter invertebrates to identify all the other invertebrates shown in the photographs on page 86.

DON'T FORGET

You must begin at the 1st pair of statements and go to wherever you are directed, for each organism you try to identify.

ONLINE

Explore keys for identifying organisms further by following the links at www.brightredbooks.net

DON'T FORGET

If you use a particular feature to separate some of the organisms involved, you can still use the same feature again to separate the other organisms.

DISTRIBUTION OF ORGANISMS 5

FACTORS AFFECTING BIODIVERSITY

Biodiversity refers to the range of different **species** present in an **ecosystem**. It also includes the different variations that exist within each species. Therefore, biodiversity covers the whole range of genetic information present in any environment.

Biodiversity can refer to an individual ecosystem, but it is sometimes used with reference to larger systems such as the whole planet. The greater the biodiversity, the healthier and more stable the ecosystem.

Various factors can affect biodiversity.

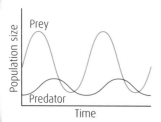

Soil erosion due to overgrazing

THE EFFECT OF BIOTIC FACTORS ON BIODIVERSITY

Biotic factors involve living organisms. There are many ways in which organisms can affect other species in an ecosystem. If the effect is extreme, it may have an impact on the biodiversity of the ecosystem.

Grazing

Grazing animals can reduce or increase biodiversity:

- Overgrazing – in severe cases, overgrazing can result in soil erosion, causing a loss of plant species from the land.
- A less drastic effect of overgrazing may be the loss of some of the less abundant plant species. As the grazing animals compete for diminishing food sources, such plant species may be eaten out of existence.
- Undergrazing – this can also reduce plant biodiversity. Vigorous plant species are able to grow unchecked and are able to dominate their surroundings at the expense of less vigorous species.
- Moderate grazing – this can maintain or increase plant biodiversity in grasslands. This is because vigorous plant species will be eaten by the grazing animals and they will not become too dominant. Less vigorous species will be able to survive and spread in the area.

Predation

Predation is the feeding of one organism (the predator) on another organism (the prey). There are many different types of feeding relationships which may be described as predation, but in true predation the predator kills and eats the prey.

Predators reduce the numbers of their prey and in extreme situations this can lead to the loss of a prey species. However, in a stable ecosystem predation results in fluctuations in the numbers of both the predator and the prey species in such a way that both species survive. The predators are obviously dependent on the prey for their survival, but the prey species also needs the predators. Without the predators, the prey **population** would increase to a point where its food supply would be completely used up. This could cause widespread starvation.

This can be seen in graphs which plot the populations of a predator species and its prey.

When the prey population is high, the predator population increases because the food supply is plentiful. As predator numbers increase, there will be increased predation and a resulting fall in prey numbers. The decrease in the prey population means less food for the predators and so their numbers will fall. The prey population will then start to increase, and so on.

A typical predator–prey relationship

THE EFFECT OF ABIOTIC FACTORS ON BIODIVERSITY

Abiotic factors do not involve living organisms. They are the chemical and physical components of an ecosystem that can determine which species are able to survive there.

- Chemical components include oxygen concentration, soil mineral concentration and water availability.

- Physical components include light intensity, pH and temperature.

pH

pH is a measure of the acidity or alkalinity of liquids. Most natural aquatic environments have a pH value in the range 6–8, which is around neutral. Most fish are adapted to live in this pH range.

Many rivers, lakes and oceans are showing signs of acidification due to **pollution**. The changes are small because there are natural substances which buffer acidic pollutants, reducing their impact. However, long-term changes in pH will affect the distribution of fish in affected waters.

Some fish are adapted to live in very extreme pH conditions:

- The black piranha lives in the Rio Negra, which is a tributary of the Amazon River in Brazil. The waters there are acidic with a pH of 3·5 to 4·5.

- The Magadi tilapia lives in Lake Magadi in Kenya. The lake is alkaline with a pH of 10.

Temperature

Oceans are undergoing rising temperatures as a result of climate change. This is having an effect on the distribution of some fish species.

In the North Sea some species such as sardines and anchovies, which are adapted to cold waters, are reported to be moving further north as water temperatures increase. At the same time, species which are adapted to warmer waters, such as red mullet and sea bass, are moving into the North Sea.

Extreme survivor: black piranha

THINGS TO DO AND THINK ABOUT

The pictures show the pellets of an owl, which is a predator. Predation is a biotic factor affecting biodiversity.

Owls regurgitate pellets which contain the undigested parts of the animals they have eaten. If the pellets are separated, the bones they contain can be used to identify the prey of the owl.

Can you think of any other examples of biotic and abiotic factors which affect biodiversity?

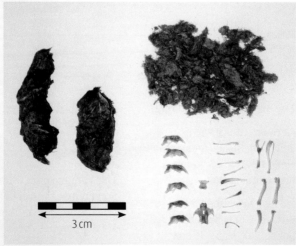
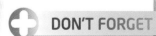

3 cm

An owl pellet   **Contents of a pellet**

DISTRIBUTION OF ORGANISMS 6

INDICATOR SPECIES

Changes to **habitats** occur naturally as a result of changing environmental conditions. Natural changes normally take place slowly and organisms may adapt or evolve with the changes. Humans have been responsible for altering their environment for thousands of years and these changes have always had an impact on biodiversity:

- Agricultural development has involved clearing forests and woodlands to make room for the cultivation of crops. Keeping herds of grazing animals has involved the removal of other grazers which would compete with domesticated cattle and sheep. It has also led to the elimination of natural predators such as wolves.

- The development of large towns and cities has reduced biodiversity in the areas concerned.

- The rapidly increasing human population means that these pressures on the environment are increasing.

- Indicator species are species that, by their presence or absence, indicate environmental quality. Therefore they are important in monitoring levels of pollution and its effects on biodiversity in an ecosystem.

VIDEO LINK

Learn more about identifying pollution at www. brightredbooks.net

AIR POLLUTION

Some of the main pollutants of the atmosphere are:

- sulphur dioxide and oxides of nitrogen – these gases come from the burning of fossil fuels, such as coal and oil, and they combine with moisture in the air to form acids

- carbon dioxide and methane – the burning of many fuels produces carbon dioxide gas; methane is released from a number of sources including sewage treatment, cattle breath and landfill sites.

Lichens are simple organisms. Each species of lichen consists of a type of fungus growing with a type of algae. They often grow on tree branches and they are so sensitive to air pollution, particularly to sulphur dioxide, that they are used as **indicator species** for air pollution. In heavily polluted air, no lichens grow on tree branches.

DON'T FORGET

Carbon dioxide and methane are known as greenhouse gases. This is because of their effect in preventing the escape of heat from the earth's atmosphere. By doing this, both gases are contributing to global warming.

Lichens as indicators of air quality

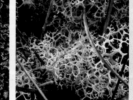

| *Crusty lichen* | *Foliose lichen* | *Fruticose lichen* | *Lung lichen* |

Polluted air ⟶ Exceptionally clean air

WATER POLLUTION

Water pollution can be caused by many things including:

- Untreated sewage entering a river – the sewage provides a food source for microorganisms and so their numbers increase in the water. They use up so much oxygen that larger organisms cannot survive.

- Agricultural **fertilisers** – these may be washed into water from neighbouring fields. They may produce a great increase in the growth of aquatic algae. These

eventually die and decay, causing similar effects to those of untreated sewage.

- Acid rain – sulphur dioxide and nitrogen oxides can dissolve in atmospheric moisture and fall as acid rain. This can make the water in rivers and lakes more acidic, preventing some species from surviving. Mussels are used as an indicator species to monitor coastal waters. Mayfly nymphs are used for freshwater.

contd

Other examples of indicator species are:

EXAMPLE 1   Nymphs of mayflies and stoneflies

These are the juvenile forms of flying insects. They spend several years in freshwater rivers and streams before emerging as flying adults. They can only survive in unpolluted water which has a high dissolved oxygen concentration and so they are used as indicators of little or no pollution.

VIDEO LINK

For more, check out the 'Water Pollution' link at www.brightredbooks.net/N5Biology

Mayfly nymph

Stonefly nymph

EXAMPLE 2   Tubifex worms and rat-tailed maggots

Tubifex worms are small segmented worms, sometimes known as sludge worms. Rat-tailed maggots are the larval stages of drone flies. They have long breathing tubes through which they can breathe air. Both species can survive in polluted, oxygen-deficient water and they are used as indicators of these conditions.

Tubifex worms

Rat-tailed maggot

BIOLOGICAL CONTROL

Biological control is the use of natural predators and diseases to control pest organisms as an alternative to the use of toxic chemicals.

EXAMPLE 1

The use of predatory Ladybird beetles to eat aphid pests.

Ladybird eating an aphid

EXAMPLE 2

The use of virus disease myxomatosis to reduce the population of rabbits, which were a pest in Australia.

Rabbit affected by myxomatosis

 THINGS TO DO AND THINK ABOUT

Blackspot is a fungal disease of rose plants. Suggest why the disease is more common in rural areas than in urban areas.

PHOTOSYNTHESIS 1

Photosynthesis is a two-stage process. The first part of the reaction is dependent on the presence of light. The second part is not dependent on light, but it does require a continuous supply of the products from the light stage. The second stage is, however, temperature dependent.

STAGE ONE: THE LIGHT REACTIONS

The first stage of photosynthesis is often known as the **light reactions** or **photolysis**. The light energy is trapped by chlorophyll in the chloroplasts of cells and is converted into chemical energy, stored in molecules of ATP. This happens in a series of steps as shown in the diagram below.

Water (H_2O) consists of molecules of hydrogen and oxygen. The light trapped by the chloroplasts provides enough energy to split the hydrogen from the oxygen. The name of this process, photolysis, explains what happens: 'photo' means that light is involved, and 'lysis' means something is split.

The oxygen which results from this process is a by-product and is not required for the reaction to proceed. It is released from the cell, diffuses into the moist air spaces of the spongy mesophyll and then passes out of the leaf into the atmosphere.

The hydrogen, however, plays an essential role in the second stage of photosynthesis and must be transferred to that part of the reaction.

Not all of the energy trapped by the chlorophyll gets used up in the splitting of water. Some of the energy is used to form ATP (adenosine triphosphate). This 'stores' the energy until it is required in stage two of the photosynthesis reaction.

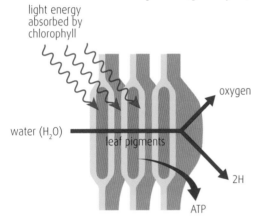

Light stage of photosynthesis

STAGE TWO: CARBON FIXATION

This stage is known as **carbon fixation** and does not require light to proceed. The energy used to drive this reaction comes from the ATP molecules which were formed in the light reactions.

As with stage one (the light reaction) this second reaction also takes place in the chloroplasts of plant cells. Several different enzymes are involved and they control a series of reactions which take the form of a cycle. Although this stage of photosynthesis is not light dependent, it is temperature dependent; enzymes must be at their optimum temperature if they are to work efficiently.

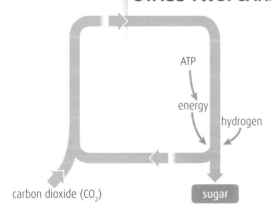

Carbon fixation stage of photosynthesis

contd

The raw material, carbon dioxide, goes through a series of reactions in which it eventually combines with the hydrogen from the light stage of the reaction. This reduces the carbon dioxide to carbohydrate. The energy to do this comes from the ATP from the light stage. ATP is broken down, releasing the energy 'stored' there.

The carbon from the carbon dioxide becomes 'fixed' into carbohydrate when it combines with the hydrogen. Carbon fixation results in the formation of sugar, often in the form of glucose.

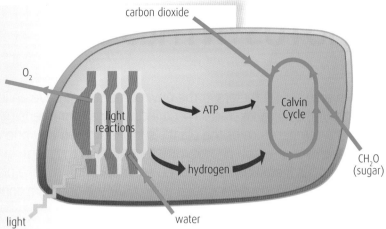

Summary of photosynthesis reactions inside a chloroplast

The process of photosynthesis can be summarised by a word summary:

$$\text{carbon dioxide + water} \xrightarrow{\text{light energy}} \text{sugar + oxygen}$$

THE FATE OF CARBOHYDRATE PRODUCED IN PHOTOSYNTHESIS

The sugar formed may be used immediately as a food in the cell in which it is produced. In this case, it would be involved in the process of respiration which is explained on pages 24 and 25. However, if not required for energy straightaway, it can be converted into other compounds.

Some of the sugar may be converted into cellulose for the construction of plant cell walls.

In many plants, excess sugar is converted into starch for storage. As starch is insoluble, it does not cause problems for the cell in terms of osmosis.

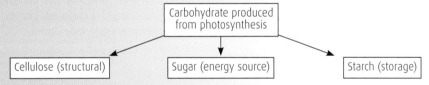

A leaf can be tested for the presence of starch as evidence that photosynthesis has taken place.

Iodine solution turns blue–black in contact with starch. However, leaves must be boiled in water to kill the cells and then be boiled in alcohol to dissolve the chlorophyll before the addition of the iodine solution.

THINGS TO DO AND THINK ABOUT

1 Chloroplasts consist of two main areas. One part is made up of stacks of photosynthetic pigments such as chlorophyll and the other part is a liquid which contains a great variety of enzymes. Can you identify in which of these parts of the chloroplast the light reaction must take place? Explain your answer.

2 What two substances from the light stage of photosynthesis are used in the carbon fixation stage?

3 What type of energy is light converted into during photosynthesis?

4 What are the chemical elements that are present in the final product of photosynthesis?

5 What is the by-product of photosynthesis and what happens to it when it is produced?

 DON'T FORGET

The first stage of photosynthesis requires light as an energy source and cannot take place in the dark. The second stage of photosynthesis uses the energy from the ATP which was formed in the first stage, so this stage can take place whether it is light or dark. However, in darkness, the supply of ATP and hydrogen soon runs out and so even the light independent stage stops.

 VIDEO LINK

For a video explanation of this experiment, watch the clip at www.brightredbooks. net/N5Biology

ONLINE TEST

For a test on photosynthesis, visit www.brightredbooks. net/N5Biology

PHOTOSYNTHESIS 2

There are several factors in the environment that have an influence on the rate at which photosynthesis takes place. These include the light intensity, the concentration of carbon dioxide and the temperature of the plant's surroundings. If any one of these factors is insufficient, it will limit the rate at which photosynthesis occurs in the plant. In turn, this affects the growth rate of the plant, as less sugar is produced for respiration to take place. These factors are important in crop production – growers need to provide the optimum conditions for maximum growth.

MEASURING THE RATE OF PHOTOSYNTHESIS

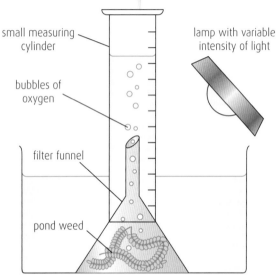

small measuring cylinder

lamp with variable intensity of light

bubbles of oxygen

filter funnel

pond weed

Pond weed bubble experiment

There are various ways to measure the rate of photosynthesis in a plant. The mass of carbohydrate produced can be calculated, but the plant may be destroyed to get the measurement. Measuring the rate of uptake of carbon dioxide or the rate of production of oxygen are two methods which can be carried out without damage to the plant.

The diagram shows the apparatus which can be used to measure the rate of oxygen production.

By altering various environmental factors, this arrangement can be used to investigate how each affects the rate of photosynthesis.

The apparatus should be set up as shown and left to adjust to the environmental conditions for a period of time. Then, the number of bubbles of oxygen produced per minute from the cut ends of the pond weed is counted. This is repeated several times and an average calculated.

Measuring the effect of light intensity

To investigate the effect of light intensity on the rate of photosynthesis, the procedure is repeated at different light intensities. This can be done using a variable intensity lamp or by moving the lamp different distances away from the plant.

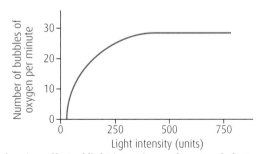

Graph showing effect of light intensity on the rate of photosynthesis

A graph of a set of results obtained from this investigation shows that, as the light intensity increases, the rate of photosynthesis (as shown by the increase in the production of oxygen bubbles) also increases – until it reaches a point at which it stops increasing. It then continues at a constant level. Further increase in light intensity has no effect because some other factor is now preventing the reaction from going faster. This factor, which is now in short supply, is called a **limiting factor**.

Measuring the effect of carbon dioxide concentration

To investigate the effect of carbon dioxide concentration on the rate of photosynthesis, the procedure is repeated at different CO_2 concentrations. This can be done by adding different quantities of sodium hydrogen carbonate to the water surrounding the pondweed. All other factors are kept at a constant level.

A graph of a set of results obtained from this investigation shows that, as the carbon dioxide concentration increases, the rate of photosynthesis also increases. At the start,

contd

it is the level of carbon dioxide which is the limiting factor. If given more CO_2, the rate of photosynthesis again increases, until it reaches a point where it stops increasing and continues at a constant level. Further increase in CO_2 concentration has no effect because again, some other factor is now preventing the reaction from going faster. The graph is a similar shape to the one for light intensity.

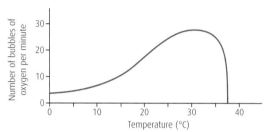

Graph showing effect of carbon dioxide concentration on the rate of photosynthesis

Measuring the effect of temperature

To investigate the effect of temperature on the rate of photosynthesis, again the same apparatus can be adapted. This time the light intensity and CO_2 concentrations are kept at a constant level and the whole apparatus is placed in a large temperature-controlled water bath.

A graph of a set of results obtained from this investigation shows a different pattern from the previous experiments. As the temperature increases, the rate of photosynthesis increases until it reaches its optimum. This is likely to be around 25–35°C, depending on the species of plant. Beyond the optimum temperature, the rate of photosynthesis falls sharply. This is due to the fact that photosynthesis is a series of enzyme-controlled reactions. As the temperature becomes higher, the enzymes are denatured and can no longer catalyse the reactions.

Graph showing effect of temperature on the rate of photosynthesis

FACTORS LIMITING PHOTOSYNTHESIS

All of the above graphs show results from investigations into how each of the three factors affect the rate of photosynthesis. However, at any one time, only one of the factors will be determining the rate. Any factor that is in short supply is known as a limiting factor.

The following graph shows the results of an *Elodea* bubbler experiment to investigate the effect of increasing light intensity, but carried out at two different temperatures. Most plants experience a range of temperatures and light intensities on a daily basis.

At point A on the graph, the limiting factor is light intensity, as increasing the light brings about an increase in the rate of photosynthesis. However, where the graph levels out, a further increase in the light intensity has no effect, showing that something else is limiting the rate of reaction.

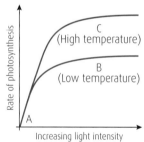

Graph showing effect of increasing light intensity at two temperatures

By comparing the graph lines at points B and C, the limiting factor is shown to be temperature – the rate of photosynthesis increases when the temperature is raised.

THINGS TO DO AND THINK ABOUT

1   A variegated plant has green and white areas forming a pattern on its leaves. Compare the rate of photosynthesis in a variegated plant to a normal green plant. Give a reason for your answer.

2   Explain why the rate of the light reaction in photosynthesis is not affected by temperature, while carbon fixation is affected.

3   Why is an aquatic plant used to measure the rate of photosynthesis rather than a land plant?

ENERGY IN ECOSYSTEMS

In food chains and food webs, the transfer of energy from one organism to the next is not complete. This is because some energy is used for heat or movement and is lost from the food chain. Some energy remains in undigested material and so is not passed to the next organism in the chain. Only a small proportion of the energy from one organism is used for the growth of the organism which eats it. Only this energy will be available to be passed on to the next stage of the food chain.

PYRAMIDS

Pyramid-shaped diagrams of various sorts are used to represent some of the features of a food chain. Here are some examples.

Pyramid of numbers

The width of each level of a **pyramid of numbers** represents the relative numbers of each organism present in the ecosystem. Therefore, the wider the level, the greater the number of organisms. For example, the diagram represents a typical food chain.

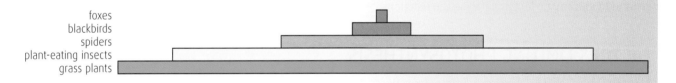

foxes
blackbirds
spiders
plant-eating insects
grass plants

However, some food chains produce irregular shapes when represented by pyramids of numbers.

For example, look at the following food chains and their representative pyramids.

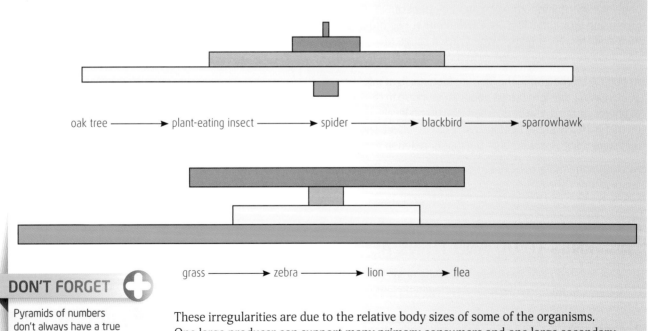

oak tree ⟶ plant-eating insect ⟶ spider ⟶ blackbird ⟶ sparrowhawk

grass ⟶ zebra ⟶ lion ⟶ flea

DON'T FORGET

Pyramids of numbers don't always have a true pyramid shape.

These irregularities are due to the relative body sizes of some of the organisms. One large producer can support many primary consumers and one large secondary consumer can support many small **parasites** (organisms which live on or in the bodies of other living things).

contd

Pyramid of energy

The best representation of the organisms present in a food chain is to estimate the amount of energy each species produces per square metre in one year (kJ/m²/year). The energy value is calculated from the dry mass involved.

Here is a **pyramid of energy** for the same food chain.

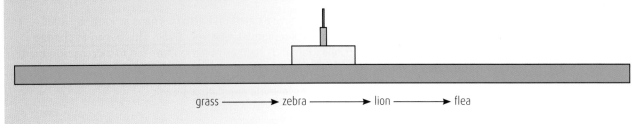

grass ⟶ zebra ⟶ lion ⟶ flea

ONLINE TEST

Ready to take the 'Pyramids' test? Logon and try at www.brightredbooks.net/N5Biology

THINGS TO DO AND THINK ABOUT

1 Explain why the number of organisms tend to decrease at successive levels of a food chain.

2 Three food chains are shown below:

(a) oak tree → greenfly → ladybird beetle → thrush → sparrowhawk

(b) grass → grasshopper → frog → fox → flea

(c) grass → grasshopper → frog → snake → eagle

Identify the food chain which corresponds to each of the following pyramids of numbers.

ONLINE

Learn more about pyramids of energy by following the link at www.brightredbooks.net/N5Biology

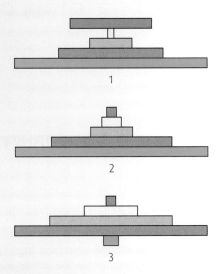

3 Explain why a pyramid of energy always has a true pyramid shape.

FOOD PRODUCTION

THE INCREASING HUMAN POPULATION

For most of the time that humans have existed on Earth (perhaps as long as 50 000 years), the total population size has remained relatively small. This began to change in the 1800s when industrialisation and changes in agriculture led to increases in food production and wealth.

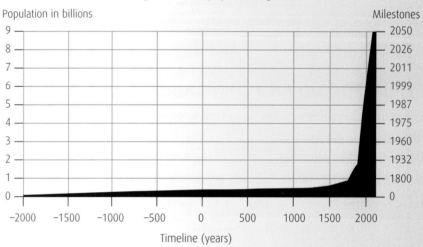

Changes in world population growth

Notice that the total population did not reach 1 billion until about 1800. It then took only another 132 years to reach 2 billion, then another 28 years to reach 3 billion, 15 more years to reach 4 billion, 12 years to reach 5 billion and 12 more to reach 6 billion. At the present time, there are more than 7 billion humans alive.

It is possible to produce enough food to feed all of the Earth's population but problems of distribution, conflicts, political disputes and adverse weather conditions mean that widespread starvation still occurs in some parts of the world.

DON'T FORGET

An increase in food production is necessary to support the world's increasing human population. This inevitably has effects on the Earth's ecosystems and on other species of animals and plants.

METHODS OF INCREASING FOOD PRODUCTION

The dramatic increase in the human population has been possible because of increases in food production. This has been due to a number of factors:

- Increased mechanisation – this has led to more intensive farming, which gives an increase in production using fewer workers.

- **Monoculture** – this is the growing of a single crop on a large scale. Fewer types of resources are needed, such as the machinery for soil preparation, planting, harvesting, storage and transport. Monoculture encourages specific pests because of the abundance of a particular crop.

- **Pesticides** – crops can be badly affected by a range of pests. These include insects which feed on the crop plants, weeds which compete for resources needed by the crop plants and fungi and viruses which cause diseases in the crop plants. Pesticides are chemicals which are used to prevent loss of crops by killing the pests.

- Genetically modified (GM) crops – it is now possible to insert specific genes into varieties of crop plants to give them desirable characteristics, such as increased growth rates, resistance to pests and diseases, and increased content of specific

contd

nutrients. The use of GM food is subject to legal approval and this varies from country to country. Early use of GM crops was in the production of vegetable oil used for cooking and margarine manufacture, and in the production of sugar and corn. The use of GM crops is expected to increase as time goes on.

- Fertilisers – planting and harvesting crops results in the continual removal of nutrients from the soil. When crops are harvested, these nutrients are not replaced by the natural processes of death and decay. The application of manure has long been used to replace nutrients in the soil, but in modern agriculture artificial fertilisers are used as the main method of replacing important minerals.

Nitrates dissolve in soil water and are able to be absorbed by plants through their roots. In plants they are used to make amino acids which can be synthesised into protein molecules. These pass from plants to animals through the food chains. The death and decomposition of organisms may return nitrogen back into the soil.

Many other important elements are recycled in a similar way.

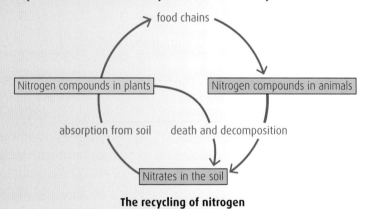

The recycling of nitrogen

ONLINE TEST

Check how much you know about intensive food production at www. brightredbooks.net/ N5Biology

PROBLEMS ASSOCIATED WITH AGRICULTURAL CHEMICALS

- Fertilisers - Sometimes fertilisers are washed out of the soil and into waterways such as rivers and lochs. This can lead to increases in the growth of aquatic algae, called an **algal bloom**, which reduces light levels in the water. This can lead to the death of aquatic plants. Dead plants and dead algae are food for bacteria in the water. Numbers of bacteria increase and use up dissolved oxygen from the water. This can lead to the death of other organisms such as fish and aquatic invertebrates.

- Pesticides – The chemicals used to control the pests of crops and of livestock are, by definition, toxic. In the past some of these chemicals were found to be persistent. This means that they remain in the tissues of organisms. The amount of such a chemical in the body of a single pest organism may be small but the amounts increase in the bodies of other organisms which eat the pests. Therefore levels of toxic chemicals can increase along a food chain. This is called **bioaccumulation**. Populations of many birds of prey declined because of the use of persistent pesticides such as DDT. Its use as an agricultural pesticide is now banned but it is still used in some parts of the world to control insect pests which transmit diseases such as malaria. Despite being banned in the USA in 1972, DDT was found in almost all human blood samples tested there in 2005. DDT was not banned in Britain until 1984. Alternative methods of pest control, such as biological control and the genetic modification of crops to make them resistant to pests, can help to reduce the reliance on pesticides.

ONLINE

Find out more about the effects of intense farming online at www. brightredbooks.net/ N5Biology

 THINGS TO DO AND THINK ABOUT

DDT has been found to be present in human milk and in the tissues of ocean fish.

How can this be explained?

EVOLUTION OF SPECIES 1

MUTATION OF GENETIC INFORMATION

The characteristics of organisms are controlled by the genes located on chromosomes. The main component of a chromosome is a molecule of DNA, arranged as a double helix. Each chromosome contains many genes. Each gene consists of a length of the DNA which has the base sequence code for the synthesis of one polypeptide. Each gene may exist in a number of different forms, called **alleles**.

Body cells are diploid; that is they contain two complete sets of chromosomes. This means that there are two alleles present for each characteristic.

Gametes are haploid and contain only one complete set of chromosomes. This means that they carry only one allele for each characteristic.

Changes to genetic information are called **mutations**. They occur as the result of random errors during cell divisions.

Mutations can occur during the division of body cells. These may have adverse effects, such as the development of cancers, but they have no evolutionary significance because they cannot be inherited.

Mutations which occur during the production of gametes can be inherited, if such a gamete is involved in fertilisation. In this case, the mutation forms part of the genotype of the new individual and may be passed on to their offspring.

Some mutations are the result of damage to the structure of chromosomes, leading to the loss of genes or to the loss of gene function. Other mutations result from errors in the DNA base-sequence copying process. These mutations change the genetic information and are the only way in which new genetic information can arise naturally. In other words, they are the only source of new alleles that can alter the characteristics of a species. The ability of a species to adapt and to evolve depends on changes to the genetic information.

ONLINE

To find out more about how DNA mutation works, visit www.brightredbooks.net/N5Biology

EFFECTS OF MUTATION

Mutations affect the genetic information present in the gametes of an organism.

Disadvantageous mutations

The altered genetic information gives rise to a protein that is unlikely to function normally. The individual organism which inherits this information may suffer because of this. Such mutations are described as disadvantageous.

> **EXAMPLE:**
>
> An example is a mutation of the gene which codes for the production of a blood-clotting protein. Lack of this protein causes haemophilia, a disorder that reduces the ability of the blood to clot. It is estimated that 33% of cases of haemophilia are the result of spontaneous mutation. The remaining cases are the result of inheriting the defective allele.

Neutral mutations

Some mutations cause no adverse consequences and are described as neutral.

> **EXAMPLE:**
>
> Some different DNA base sequences code for the same amino acid and so some mutations which alter the DNA code have no effect on the resulting proteins.

VIDEO LINK

Go online and watch the video to learn more about genetic mutation at www.brightredbooks.net/N5Biology

contd

Advantageous mutations

In some very rare instances, a mutation may cause a genetic change producing a new allele which has a beneficial effect. In this case, the mutation is described as advantageous.

> **EXAMPLE:**
>
> An example of this is a mutation of a gene which codes for a protein that helps to prevent accumulation of cholesterol in arteries. The resulting protein is even better than the normal protein at carrying out this function. People possessing this mutation have a reduced risk of heart disease compared to the rest of the population. About 3·5% of people in a village in northern Italy possess this gene. It can be traced back to a mutation in one man who lived there in the eighteenth century and who passed it on to his offspring.

CAUSES OF MUTATION

Mutations are random. They happen spontaneously and cannot be predicted. However, it is known that certain environmental factors can increase the rate at which mutations occur. This is because they increase the chance of errors during the formation of gametes.

Such environmental factors are called **mutagens** and they include various forms of radiation, for example ultraviolet light, X-rays and gamma radiation. Some chemicals are also mutagens, for example asbestos fibres, benzene, mustard gas and tobacco smoke.

ONLINE TEST

Test your knowledge of mutation and its causes at www.brightredbooks.net/N5Biology

DON'T FORGET

Remember a mutation is a change to the genetic information in an organism. It is not the organism itself!

 THINGS TO DO AND THINK ABOUT

This type of apron has a lead lining. It is used by staff who operate X-ray machines.

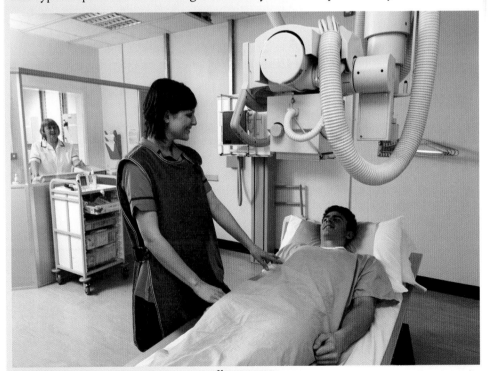

X-ray apron

1 Why are such precautions necessary?

2 Why is it especially important that such clothing protects the reproductive organs?

EVOLUTION OF SPECIES 2

ADAPTATION AND VARIATION

An **adaptation** is an inherited characteristic that makes an organism well suited to its environment, enabling it to survive. Adaptations are the result of variations which appear in a population of organisms due to random advantageous mutations. Such variations may allow a species to adapt to changing environmental conditions, or to become better adapted to the existing conditions.

EXAMPLES OF ADVANTAGEOUS ADAPTATION

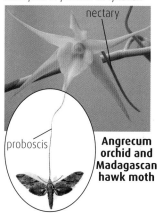

Kangaroo rats are extremely well adapted to desert living

Examples of parallel adaptation

nectary

proboscis

Angrecum orchid and Madagascan hawk moth

EXAMPLE 1   Kangaroo rats

These are desert mammals that are able to survive in regions of high temperatures and low water availability. They show many adaptations, including: high levels of a hormone that promotes reabsorption of water from the kidneys back into the blood; remaining in cool humid burrows during the day; coming out to feed at night; producing very dry faeces; having no sweat glands.

EXAMPLE 2   Angrecum orchid and Madagascan hawk moths

The flowers of this orchid have nectaries that are 30 cm long. The proboscis (feeding tube) of the moth is equally long. The moth is the only insect able to reach the nectar and so it is the only insect capable of pollinating the flowers.

Both species benefit from this relationship. The orchid is certain to have its flowers pollinated because it is the only source of food for the moth. The moths have no competition from other species for the nectar.

This is an example of adaptations developing in two species in parallel. The result is that the two species become dependent on each other.

EXAMPLE 3   Swollen-thorn acacia and acacia ants

Swollen-thorn acacia and acacia ants

Acacia trees possess very sharp thorns and poisonous leaves which help to protect them from many grazing animals and insect pests. The swollen-thorn acacia, however, is different. Its thorns are larger and hollow, and its leaves are not poisonous. The thorns are used by one species of ant as nesting sites. The ants also feed on nectar from the flowers of the tree. In addition, the tree produces parcels of food material on their leaves which the ants collect and feed to their young. These parcels have no other function.

In return for shelter and food, the ants protect the acacia. If any animal begins grazing on the tree or any insect begins eating its leaves, the ants swarm out of their nests and attack the offender. The ants will even kill any other plants, such as vines, which try to grow up the acacia.

This is another example of adaptations developing in parallel in two species in such a way that they become dependent on each other.

The bee orchid flower

EXAMPLE 4   Bee orchid

The flowers of this plant have become adapted in such a way that they mimic a female bee. This attracts male bees to the flowers in an attempt to mate. In doing so, the bees act as pollinators for the orchid.

contd

These examples demonstrate that adaptation is a continuous process for all species. If a species does not show continuous adaptation, it will fail to survive because it will eventually be out-competed by another species.

This is sometimes referred to as the Red Queen model of evolutionary change. It is a reference to the Red Queen in Lewis Carroll's book *Through The Looking Glass* in which the Red Queen states,

'... here it takes all the running you can do just to remain in the same place.'

EXAMPLES OF ADAPTATIONS WHICH CAUSE PROBLEMS

Although adaptation is largely advantageous to the survival of species, sometimes it can cause problems. Here are two examples.

EXAMPLE 1   Over-prescription of antibiotics

Antibiotics are powerful tools in combating diseases caused by bacterial infections. They kill disease-causing bacteria. Unfortunately, bacteria have the ability to produce variants, some of which may be resistant to the antibiotic that is being used against them. This resistance can be passed on when the bacteria reproduce. Even if just a few bacteria survive the antibiotic treatment, they can quickly produce a population of offspring that are all resistant. This means that an alternative antibiotic, or a stronger dose of the original antibiotic, would be needed in future treatments. Continued variation in the bacteria eventually may create resistance to the new or stronger antibiotic.

The over-prescription of antibiotics for the treatment of minor infections has been one of the reasons for the development of resistant strains of bacteria. This process has led to some infections being very difficult to treat.

For example, MRSA (methicillin-resistant *Staphylococcus aureus*) and C. diff. (*Clostridium difficile*) are bacteria which cause great problems in some hospitals because of their resistance to normal antibiotics and because of the danger they pose to patients in a weakened condition.

EXAMPLE 2   Pests of GM crops

Genetic modification has been used to give some crops in-built protection against serious pests by inserting genes which give the plants the ability to produce toxins. For example, some varieties of cotton have been modified to make them resistant to the caterpillars of the bollworm moth. Similarly, some varieties of maize have been modified to make them resistant to the caterpillars of the European corn borer moth. We call these **GM organisms**.

In both cases, it is the caterpillars of the moths which cause the damage to the crops and, in both cases, adaptations arise which give the caterpillars resistance to the toxins produced by the crop plants. Potentially, this could lead to problems of widespread resistance in the moths. To reduce the risk of this happening, the GM crops are grown alongside non-GM varieties of the crop. This reduces the overall damage caused by the moths, while not increasing the chance that **natural selection** will produce populations of moths that are all resistant to the plant toxins.

THINGS TO DO AND THINK ABOUT

1   For each of the adaptations of the kangaroo rat described in Example 1, decide whether it is a behavioural or a physiological adaptation and say how it contributes to the survival of the animal.

2   How does the Lewis Carroll quote of the Red Queen illustrate the idea of the need for constant adaptation by a species?

3   What are the potential benefits and potential problems which can result from the use of genetic modification of organisms in the production of foods?

DON'T FORGET

Adaptation can occur in parallel between two species or singularly in one species and is the result of random mutations.

ONLINE TEST

For a test on adaptation, visit www.brightredbooks.net/N5Biology

ONLINE

To learn more about the over-prescription of antibiotics and the problems it can cause, read the NHS article on the subject at www.brightredbooks.net/N5Biology

EVOLUTION OF SPECIES 3

NATURAL SELECTION AND SPECIATION

Natural selection

Natural selection is sometimes referred to as **'the survival of the fittest'**. It is the result of two factors acting together.

- In normal conditions, every species produces more offspring than the environment can support. This produces intraspecific competition between the individuals of a species.

- Variations occur naturally in a species. This means that some individuals will be better adapted to their environment than others. These individuals have a greater chance of surviving, reproducing and passing on the genetic information which gave them the beneficial variations.

In this way, the environment, or nature, selects individuals who possess beneficial variations. These individuals survive better than individuals who do not possess the variations. In the species as a whole, the genes which produce the beneficial variations persist and increase in the population. Genes that produce less beneficial or disadvantageous variations become less frequent in the population.

Speciation

A species is a group of organisms which can interbreed to produce fertile offspring.

Speciation is the evolution of new species. It happens if part of a population of organisms becomes isolated from the rest of the population by a barrier. Natural selection will continue in both the main population and in the isolated group. Selection pressures may differ for each group, favouring different variations. Given enough time, the two groups may become genetically different to the extent that they are considered to be different species.

Isolating barriers can be:

- geographical – for example mountains, deserts, rivers and oceans

- ecological – for example occupying different habitats in the same geographical area or abiotic factors such as pH

- behavioural – for example differences in courtship behaviour or the failure of sex cells to fuse.

Here are some examples of speciation.

EXAMPLE 1 Galapagos finches

The Galapagos Islands are volcanic islands in the Pacific Ocean. In the years following their formation they were gradually colonised by plants and animals which arrived there by chance and which became isolated from the mainland populations.

Included in these arrivals were members of a species of finch from the mainland of South America. Through the mechanism of variation and natural selection, this single species of finch has evolved into a range of different species which utilise different food sources. In this way, they occupy different ecological niches. This was not possible on the mainland because the equivalent niches were already occupied by other species. Niches are covered on page 74.

The Galapagos finches evolved from a single species from mainland South America

contd

EXAMPLE 2   Arctic char

The Arctic char is a species of fish related to salmon. Like salmon, they breed in freshwater and migrate to the sea as young fish. They return to rivers to breed when they are mature.

During the last ice age, populations of these fish became trapped in some deep freshwater lochs in Scotland. Here they became isolated populations and have evolved into a true freshwater species.

They are a relatively rare species in Scotland but attempts are being made to farm them for food, in the same way as salmon are farmed.

The Arctic char

EXAMPLE 3   Arran whitebeam

The Arran whitebeam is a species of tree found only on the Scottish island of Arran. It is thought to be a hybrid of two related tree species. It has become established in an ecological niche which the two parent species are unable to occupy successfully. It is found in only a few isolated sites on the island and is considered to be in danger of extinction.

The Arran whitebeam

EXAMPLE 4   St Kilda wren

St Kilda is a remote Scottish island which was inhabited until 1930, when the remaining residents were evacuated to mainland Scotland.

It has a population of wrens that is a different sub-species to the mainland population of wrens. The St Kilda wren is larger, heavier, has longer wings and a stouter bill. Given enough time, this isolated population will evolve into a separate species.

ONLINE

Find out more about the wildlife of St. Kilda by watching the clip at www.brightredbooks.net/N5Biology

ONLINE TEST

To see how much you've learned about natural selection and speciation, test yourself at www.brightredbooks.net/N5Biology

St Kilda wren

Mainland wren

THINGS TO DO AND THINK ABOUT

1   For each of the Examples 1–4 decide what barrier caused the isolation of one population from other populations of the same species.

2   What other factors could result in isolation of one population from another?

GLOSSARY

Abiotic factor
A non-living chemical or physical factor in the environment which has an effect on the ecosystem. Chemical factors include industrial waste, pH and fertiliser run-off. Physical factors include light, temperature and humidity.

Active site
The area on the surface of an enzyme molecule where substrate molecules become attached and undergo a change.

Active transport
The movement of a substance across a cell membrane from an area of lower concentration to one of higher concentration. This requires the expenditure of energy by the cell.

Adaptation
A characteristic of an organism or species that has developed by evolution and which aids its survival. It could also mean the process of adapting.

Aerobic respiration
The release of energy from foods using oxygen.

Algal bloom
The rapid growth in the population of aquatic algae in ponds and lakes caused by an increase in the concentration of inorganic nutrients. This could be due to fertiliser run-off or industrial waste. The subsequent death and decomposition of the algae can lead to the reduction in oxygen concentration of the water and the death of fish and other animals.

Allele
One of a number of alternative forms of the same gene. Diploid cells contain two alleles for each gene, one in each of the two sets of chromosomes. Haploid gametes contain only one allele for each gene.

Alveolus
A microscopic thin-walled air sac found at the end of the bronchioles in the lungs. The alveoli form the gas-exchange surfaces between the blood and the air.

Amino acids
Small molecules which are linked together in chains to form protein molecules.

Antibodies
Y-shaped protein molecules which are released from white blood cells called lymphocytes. They attach to pathogens such as bacteria and viruses. Each antibody is specific to a particular pathogen.

Aorta
The largest artery in the body. The aorta carries oxygenated blood from the left ventricle and branches from the aorta carry blood to all areas of the body.

Artery
A blood vessel with thick muscular walls which carries blood, under high pressure, away from the heart to other parts of the body.

ATP
The substance produced using the energy released during respiration. The energy can be released from ATP for cellular activities.

Atrium
A receiving chamber of the heart. The left atrium receives oxygenated blood from the lungs and the right atrium receives deoxygenated blood from other parts of the body.

Base pairs
The bonding of complementary nucleic bases in DNA. This enables the replication of chromosomes prior to cell division and the synthesis of proteins from amino acids in cells.

Bioaccumulation
The build-up of toxic substances in the tissues of an organism because the substance is persistent. The rate at which the substance is broken down or excreted by the organism is slower than the rate at which the organism absorbs the substance from its food.

Biodiversity
The range of different species present. Biodiversity is normally considered within an ecosystem. The greater the biodiversity, the greater the stability of the ecosystem.

Biological control
The use of living organisms to control pests. The control organism may act as a predator or it may cause a disease of the pest organism.

Biotic factor
A biological factor, involving a living organism, found in the environment and which has an effect on the ecosystem. Biotic factors include food sources, predation and disease.

Brain
The controlling part of the CNS. It regulates body processes and is capable of higher mental processes including thinking, reasoning and remembering. It consists of the cerebrum, cerebellum and the medulla.

Capillary
A microscopic blood vessel. Capillaries are the sites of the exchange of substances between the blood and the body cells. Blood passes from small arteries into capillaries and then into small veins.

Carbon dioxide
One of the end products of aerobic respiration and of fermentation in plant cells and yeast.

Carbon fixation
A stage of photosynthesis which does not require light. Carbon dioxide combines with hydrogen to form carbohydrate.

Carnivore
An animal which gets its energy by eating other animals.

Cell
The smallest living structure. Cells are the building blocks of multicellular organisms. They can also exist as independent unicellular organisms.

Cell equator
The mid-line of a cell at which pairs of chromatids become arranged during cell division.

Cell membrane
The membrane separating the interior of a cell from its outside environment. It consists of a double layer of phospholipids with associated proteins. It controls the movement of materials into and out of the cell.

Cell wall
The structural part of a plant cell found outside of the cell membrane. Cell walls are made mainly of cellulose. They hold the cells in place and give support to plant tissues. Bacteria and fungi have cell walls which differ in structure to those of plant cell walls.

Cellulose
A structural carbohydrate which forms the major component of plant cells.

Central nervous system
The central nervous system (CNS) consists of the brain and spinal cord. These areas receive sensory information, process it and initiate any necessary responses to it.

Cerebellum
The part of the brain which coordinates muscular movement and helps to maintain balance.

Cerebrum
The part of the brain responsible for higher mental processes including thinking, reasoning and remembering.

Chemical receptor
A protein on the surface of a structure which allows other chemicals to become attached, for example the surface of a pathogen to which an antibody becomes attached.

Chlorophyll
A green pigment found in the chloroplasts of green plants. It is responsible for the absorption of light during the first stage of photosynthesis.

Chloroplast
A membranous organelle present in the cytoplasm of cells found in the green areas of plants. Several chloroplasts may be present in a single cell. Chloroplasts contain the pigment chlorophyll and are the sites of photosynthesis.

Chromatid
One of a pair of identical DNA molecules which together form a chromosome.

Chromosome
Composed of DNA and associated proteins, chromosomes contain the genetic code of an organism. This controls the development and activities of the cell. The number of chromosomes present varies between different species.

Community
The populations of all the different species present in an ecosystem.

Companion cell
A small cell which is adjacent to a sieve tube cell in phloem. Sieve tube cells do not have a nucleus and the companion cells control their activities.

Competition – interspecific
Competition for a resource which occurs between individuals of different species.

Competition – intraspecific
Competition for a resource which occurs between individuals of the same species. This type of competition can be the more severe because the requirements of the individuals are identical.

Consumer
An organism which gets its energy by feeding on other organisms or their wastes.

Coronary artery
A branch of the aorta which supplies blood to the muscle cells of the heart walls, supplying them with nutrients and oxygen.

Cytoplasm
A jelly-like material within a cell. It is the site of many chemical reactions and the cell organelles are contained in it.

Degradation
A chemical reaction in which a larger substrate molecule is broken down into smaller product molecules.

Denaturation
Changes to an enzyme molecule affecting its shape and structure, often resulting from high temperatures or extremes of pH.

Diffusion
The movement of gaseous or dissolved molecules from an area of higher concentration to an area of lower concentration. It can result in the movement of small molecules into or out of a cell without any energy expenditure by the cell.

Diploid number
The number of chromosomes present in the body cells of an organism. It is represented as 2n and refers to the two sets of chromosomes inherited by the organism, one set from each parent.

DNA
Deoxyribonucleic acid – the chemical which forms the structure of chromosomes and which contains the genetic code of the organism. It exists as a long double-stranded helix.

Dominant allele
The allele which shows its effect over the other allele for a characteristic, if two different alleles are present.

Double helix
The shape of the DNA molecule in a chromosome. It refers to the two strands of the molecule which are held together by bonds between the complementary bases and twisted into a spiral shape.

Ecosystem
An ecological system consisting of a habitat together with its community of organisms. The organisms are part of a food web through which energy and nutrients are transferred.

Egg cell
The female gamete (sex cell) of animals. They contain a food store for the early development of an embryo.

Embryo
The early stage of development of an animal or plant.

GLOSSARY

Enzyme
A protein which acts as a catalyst for chemical reactions in cells. Each enzyme is specific for one type of reaction.

Epidermis
The outermost layer of cells of an organism. It acts as a barrier to protect other cells.

Ethanol
The end product (together with carbon dioxide) of fermentation in plant cells and yeast.

Evolution
The change in the inherited characteristics of organisms from generation to generation. Evolution leads to the gradual change in the characteristics of a species and also to the divergence of populations into new species.

F_1 generation
The first generation of offspring of a genetic cross (1st filial generation).

F_2 generation
The second generation of a genetic cross (2nd filial generation).

Fermentation pathway
The type of respiration which releases energy from foods in the absence of oxygen.

Fertilisation
The fusion of the nucleus of a male gamete with the nucleus of a female gamete. Each gamete is haploid (contains one set of chromosomes). The resulting zygote cell is diploid (two sets of chromosomes).

Fertiliser
Material that is added to soil to increase the supply of nutrients to plants. Fertilisers can be organic, for example manure, or inorganic, for example compounds of nitrogen, potassium and phosphorus. Excess fertiliser can cause pollution if it is washed into ponds or streams.

Food chain
A linear sequence of organisms showing the feeding relationships between them. A food chain always starts with a producer (green plant) and ends with a top predator that is not eaten by anything else. Each link in the food chain represents the transfer of energy and nutrients from one species to the next. Food chains are normally part of more complex food webs.

Food web
A complex system of interlinked food chains which represent the feeding relationships in a community of organisms. The greater the complexity of a food web then the greater is its stability. This increases its ability to adapt to any changes in the populations of individual species.

Gene
A unit of genetic information which controls a characteristic of an organism.

Genetic code
The sequence of bases on a DNA molecule which controls the sequence of amino acids in a protein.

Genetic engineering
The altering of an organism's genetic information. It can involve the transfer of genes from different species to produce improved or novel organisms.

Genotype
The genetic information possessed by an organism. It can refer to the information for a single characteristic and is usually shown as the abbreviation of the two alleles involved.

Glucagon
A hormone produced by the pancreas. It causes the conversion of glycogen to glucose by the liver. It is part of the mechanism which controls blood sugar levels.

Glucose
The sugar which is commonly used as an energy source in the process of respiration.

Glycogen
An insoluble carbohydrate consisting of chains of glucose molecules. It is used as a storage chemical in animals.

GM organism
A genetically modified organism - one that has been produced by genetic engineering.

Guard cell
A cell found mainly in the lower epidermis of plant leaves. Pairs of guard cells form a stoma which allows water vapour to pass from the leaves during transpiration.

Habitat
An ecological or environmental area in which a population of organisms live.

Haemoglobin
A protein which contains an atom of iron. Haemoglobin is found in red blood cells where it combines with oxygen for transport around the body.

Haploid number
The number of chromosomes present in the gametes of an organism. It is represented as n and refers to the single set of chromosomes.

Heart
The organ which pumps blood around the body. In mammals it consists of four chambers: two atria which receive blood, and two ventricles which pump blood.

Herbivore
An animal which gets its energy by eating plants.

Heterozygous
An organism which possesses two different alleles for a particular characteristic.

Homozygous
An organism which possesses two of the same alleles for a characteristic. It does not matter whether the alleles are both dominant or both recessive.

Hormones
Chemicals which help to control development and metabolism in the body. They are released in one part of the body and transported in the blood to act at other parts of the body. Many hormones are proteins.

Indicator species
A species of organisms whose populations are easily affected by changes in environmental conditions. Their presence or absence can indicate environmental quality or levels of pollution.

Insulin
A hormone produced by the pancreas. It causes the conversion of glucose to glycogen by the liver and the uptake of glucose by other body cells. It is part of the mechanism which controls blood sugar levels.

Isolating barrier
A mechanism which separates populations of a species from each other, allowing them to evolve separately and perhaps to become different species. Barriers can be:

- Geographical, for example rivers or mountains;
- Ecological, for example occupying different habitats in an ecosystem;
- Behavioural, for example differences in courtship behaviour.

Lactate
The end product of fermentation in animal cells and bacteria.

Lacteal
A small vessel in a villus which absorbs the products of fat digestion.

Light reactions / Photolysis
The first stage of photosynthesis in which light is absorbed by chlorophyll, leading to the production of ATP and hydrogen for use in carbon fixation.

Lignin
A substance which is deposited in the walls of xylem vessels. It strengthens the vessels for their function in water transport and eventually forms the woody tissue of trees and shrubs.

Limiting factor
An essential factor required by a process and which limits the rate of that process because it is in short supply. If several factors are required, normally only one of them will act as the limiting factor at any given time.

Liver
The organ which stores glucose as glycogen. It also breaks down old red blood cells, produces bile, breaks down harmful substances such as alcohol and breaks down excess amino acids to produce urea.

Lungs
The organs of gas exchange in mammals. The gas exchange takes place between air in numerous microscopic air sacs called alveoli and blood in adjacent blood capillaries.

Medulla
The part of the brain which connects to the spinal cord. It controls automatic activities such as the breathing rate and heart rate.

Mesophyll
The inner photosynthetic tissue of plant leaves. There are two types of mesophyll:

- Palisade mesophyll is a layer of closely packed cells near to the leaf surface. The cells contain numerous chloroplasts and carry out most of the photosynthesis;
- Spongy mesophyll is a layer of loosely packed cells below the palisade mesophyll. Air spaces between the cells allow absorption of carbon dioxide for photosynthesis and the evaporation of water in transpiration.

Messenger RNA
Messenger ribonucleic acid – a chemical similar to DNA but with smaller molecules and a single-stranded structure. It is involved in protein synthesis in cells by carrying a complementary copy of the genetic code from the DNA to the cell ribosomes.

Mitochondrion
A membranous organelle found in the cytoplasm of cells. This is the site of those stages of respiration which require oxygen.

Mitosis
The process by which the nucleus of a cell divides prior to the division of the whole cell into two daughter cells. Mitosis involves the replication of the cell's chromosomes to produce two identical diploid sets so that the daughter cells are genetically identical to the parent cell.

Monohybrid inheritance
The pattern of inheritance shown by a single characteristic. Mendel's experiments using garden peas illustrate this pattern.

Mutagen
A mutagenic agent – something that can increase the chance of a mutation taking place. For example X-rays, gamma rays and UV radiation. Some chemicals are mutagenic because they can react with DNA.

Mutation
A change to the genetic information of an organism. This can be a change to the number or structure of the chromosomes, or a change to the base sequence of a chromosome. Mutations are normally the result of mistakes during cell division. Mutations during the formation of gametes are of greater significance because they can become part of the genetic information of a new individual and be passed on to future generations.

Natural selection
The process by which variations can increase or decrease in a species because they affect the survival chances of individuals which possess them. Advantageous variations increase in the population because individuals with them have a greater chance of surviving and reproducing. Disadvantageous variations decrease in the population for the opposite reason.

Neurons
A nerve cell. Sensory neurons carry electrical impulses from sensory receptors to the CNS. Motor neurons carry electrical impulses from the CNS to effector organs such as muscles or glands. Inter neurons transfer electrical impulses within the CNS.

GLOSSARY

Niche
The role of a species in its ecosystem. The niches of different species can overlap but they are not identical. The description of a niche includes the food and other resources used by a species, its predators and the habitat it occupies.

Nucleus
The organelle which contains the chromosomes of the cell. It is surrounded by a membrane which separates the chromosomes from the cytoplasm.

Omnivore
An animal which gets its energy by feeding on both plant and animal material.

Optimum conditions
The values of factors such as temperature, pH and substrate concentration at which an enzyme-controlled reaction takes place at its maximum rate.

Organ
A structure which has a particular function, e.g. heart, stomach, leaf. An organ may be made up of one or several different tissue types.

Organelles
The various small structures present in the cytoplasm of many types of cells.

Osmosis
The movement of water molecules from an area of higher water concentration to a lower water concentration, through a selectively permeable membrane. Osmosis is a particular case of diffusion and does not require energy expenditure by a cell.

Ovary
The female reproductive organ which produces egg cells in animals and ovules in flowering plants.

Ovule
This contains the female gamete (sex cell) of flowering plants. It contains a food store for the early development of an embryo.

Oxyhaemoglobin
The form of haemoglobin produced when haemoglobin combines with oxygen.

P generation
The parental generation of a genetic cross.

Paired-statement key
An identification tool in which a series of pairs of contrasting statements are applied to an unidentified organism. Only one of the statements will be true for the organism and will lead to a further pair of statements in the key. Eventually a true statement will lead to the name of the organism.

Pancreas
The organ which produces the hormones insulin and glucagon. It also produces several digestive enzymes.

Passive transport
The movement of a substance into or out of a cell without the need for energy expenditure by the cell. Diffusion and osmosis are examples of passive transport.

Pesticide
A chemical used to kill pests. A pesticide can be specific for particular types of pests. For example herbicides kill plants, insecticides kill insects and fungicides kill fungi.

Phenotype
The characteristics possessed by an organism resulting from its inherited genetic information. Phenotypes can be modified by environmental effects.

Phloem
Plant tissue which transports dissolved sugar through the plant. Phloem consists of elongated cells which form sieve tubes and their associated companion cells.

Phospholipid
A lipid molecule important in the structure of cell membranes. Phospholipids become arranged in flexible double layered sheets which form the basic membrane structure.

Photosynthesis
The process by which green plants produce food using absorbed light energy. Carbohydrate is produced from carbon dioxide and water. Oxygen is a by-product of the process.

Pitfall trap
A sampling technique used to catch ground-living invertebrates.

Plasma
The liquid part of the blood containing dissolved substances to be transported by the blood. The red and white blood cells are carried along by the plasma.

Plasmid
A small circular molecule of DNA found in bacteria. Plasmids can be transferred between bacterial cells allowing the transfer of genetic information. This ability is used in genetic engineering.

Plasmolysed
The condition of a plant cell which has lost water by osmosis causing its vacuole to lose water and the cytoplasm and cell membrane to shrink away from the cell wall.

Pollen grain
This contains the male gamete (sex cell) of flowering plants. It is are carried to the female gametes by wind or by insects.

Polygenic
The inheritance of a characteristic which is controlled by two or more different genes. This produces greater variation of that characteristic and contributes to the continuous variation shown by some characteristics.

Population
All the members of a single species living in an area.

Potometer
An instrument used to measure the transpiration rate of a plant. A weight potometer measures the decrease in mass due to water loss from the plant. A bubble potometer measures the rate of water uptake into the cut stem of a plant.

Predator
An animal which catches and kills other animals for food.

Prey
An animal which is killed by another animal for food.

Producer

A green plant which makes its own food by photosynthesis. It is the first organism in a food chain.

Product

The substance(s) formed at the end of a chemical reaction, including enzyme-controlled reactions.

Protein

A substance made up of chains of amino acid molecules. Proteins have many functions: structural, enzymes, hormones, antibodies and chemical receptors.

Pulmonary artery

An artery which carries deoxygenated blood from the heart to the lungs.

Pulmonary vein

A vein which carries oxygenated blood from the lungs to the heart.

Punnett square

A technique used to determine the genotypes of the offspring from a cross in which the genotypes of the parents are known.

Pyramid of energy

A representation of the total energy turnover at each stage of a food chain. It gives a truer picture than a pyramid of numbers because it is scaled to equivalent units of area and time for each level of the pyramid.

Pyramid of numbers

A representation of the total numbers of organisms at each stage of a food chain. The pyramid shape is often distorted because of differences in the size of individual organisms at each stage of the pyramid.

Pyruvate

The substance produced during breakdown of glucose in the first stage of respiration.

Quadrat

A sampling tool consisting of a square frame and used to count the number of stationary organisms in a defined area.

Recessive allele

The allele which has its effect masked by the other allele for a characteristic, if two different alleles are present.

Red blood cell

A cell which transports oxygen around the body in the blood. Red blood cells are very numerous, have a biconcave shape and lack a nucleus. They contain haemoglobin which combines with oxygen to form oxyhaemoglobin.

Reflex action

An automatic involuntary response to a stimulus. Reflex actions are protective (sneezing, coughing, withdrawal from hot objects), or they help maintain body processes (saliva production, control of heart rate).

Reflex arc

A nerve pathway which controls a reflex action. It does not require a conscious input from the brain, allowing the response to be made almost instantly.

Respiration

The process by which a cell releases energy from glucose molecules. The energy is used for the production of ATP molecules. Respiration with oxygen (aerobic respiration) releases more energy from glucose than respiration without oxygen (fermentation). During aerobic respiration, glucose is broken down into carbon dioxide and water.

Respirometer

An instrument used to measure the rate of aerobic respiration in living tissue. Respirometers normally measure the rate of oxygen uptake by the tissue.

Ribosome

A small organelle which is the site of protein synthesis. There are many ribosomes in the cytoplasm of a cell.

Root hair

An extension of a plant root hair cell. Root hairs provide a large surface area for the absorption of water and minerals from the soil.

Selectively permeable

The property of cell membranes which allows small soluble molecules to pass through but which prevents larger molecules from doing so.

Sensory receptor

A cell which produces an electrical impulse in response to a stimulus. Such cells may form part of a specialised sense organ.

Sieve plate

The perforated end cell walls of two elongated sieve tube cells which allow continuous strands of cytoplasm to stretch between both cells. This allows easier movement of transported sugar along the sieve tube.

Small intestine

The region of the digestive system from which digested food is absorbed into the blood.

Speciation

The evolution of new species through the process of natural selection in isolated populations of an existing species.

Species

A group of interbreeding organisms whose offspring are fertile.

Sperm cell

The male gamete (sex cell) of animals. They are able to swim towards the female gametes.

Spinal cord

The part of the CNS which links the brain to the rest of the body. It is capable of initiating automatic reflex responses to sensory stimuli.

Spindle fibres

Microscopic fibres produced in a cell during cell division. They control the separation of chromatids between the daughter cells.

Stamen

The male reproductive organ in flowering plants. Pollen grains are produced in the anther at the tip of the stamen.

GLOSSARY

Stem cell
An undifferentiated animal cell which has the ability to divide by mitosis to produce more undifferentiated cells or to develop into one of a number of specialised cell types.

Stoma
A pore in the epidermis of a leaf. Stomata are very numerous and are formed by two curved guard cells which can open and close the pore. Most stomata are normally found on the lower surface of a leaf.

Substrate
The substance(s) to which an enzyme becomes attached and which then undergoes a chemical reaction.

Synapse
A gap between two neurons. The impulse is carried across the synapse by a chemical transmitter that is produced at the end of one neuron and detected by the other neuron.

Synthesis
A chemical reaction in which smaller substrate molecules react to form a larger product molecule.

System
A group of connected structures or organs which together have a common function. For example the digestive system, nervous system, circulatory system.

Testis
The male reproductive organ which produces sperm cells in animals.

Tissue
A group of similar cells which are arranged together and which carry out a common function. For example nerve tissue, muscle tissue, epidermal tissue.

Transect
A sampling technique involving taking samples at regular intervals along a straight line. It can be used to study the effects of changes in a variable from one area to another, for example the effect of changes in light intensity between a shaded and an exposed area.

Transpiration
The loss of water from the leaves of a plant. Water evaporates from the surfaces of spongy mesophyll cells into the air spaces of the leaf. It then diffuses through the stomata into the outside air.

Tree beating
A sampling technique used to collect organisms which live on the branches and leaves of trees. It involves shaking the tree branches so that organisms fall onto a sheet placed below the branches. It is probable that some organisms will be missed or escape.

Tullgren funnel
A sampling technique used to collect organisms from leaf litter or soil. It involves placing the sample material on a sieve in a funnel and gently heating it from above. Small organisms move downwards away from the heat and fall through the sieve into a container.

Turgid
The state of a plant cell which contains the maximum amount of water. The cell contents are swollen and push against the cell wall. This contributes to the support of the plant by making the plant tissue firm.

Vacuole
A membranous bag in the cytoplasm of a plant cell. It is filled with cell sap (a solution of sugars and salts). It is important in maintaining the shape of the cell.

Valve
Structure found in the heart and in veins to prevent the backflow of blood.

Variation - continuous
Differences in a characteristic which show a continuous range of possibilities between a minimum and a maximum value. They are normally the result of polygenic inheritance. Examples are mass and height.

Variation – discrete
Differences in a characteristic which show a limited number of distinct possibilities. They are usually the result of single-gene inheritance. Examples are blood groups and pea seed shape.

Vein
A blood vessel which carries low pressure blood towards the heart from other areas of the body.

Vena cava
The largest vein in the body. It carries deoxygenated blood from various parts of the body to the right atrium.

Ventricle
A pumping chamber of the heart. The right ventricle pumps deoxygenated blood to the lungs and the left ventricle pumps oxygenated blood to other parts of the body.

Villus
A small projection on the inner lining of the small intestine. They are very numerous and provide a large surface area for the absorption of digested food.

White blood cell
Cells found in the blood which help protect the body from infection. They are larger and less numerous than red blood cells. There are two main types of white cells:

- phagocytes engulf and digest pathogens;
- lymphocytes produce antibodies which bind onto pathogens, making them ineffective.

Xylem
Plant tissue which transports water from the roots to the leaves of a plant. It consists of elongated dead cells which form continuous hollow tubes, strengthened by lignin.

Zygote
The cell produced at fertilisation by the fusion of the nuclei of a male and female gamete. The zygote is the first cell of a new individual. It has a diploid chromosome complement which is maintained in all subsequent body cells by mitosis.